Albert Einstein's Unified Field Theory

A New Interpretation

2021 UPDATE

SUNRISE INFORMATION SERVICES

Albert Einstein's

Unified Field Theory

His Final Scientific Legacy

2021 UPDATE

SUNRISE Information Services
Canberra, Australia

A copy of this publication can be found in the National Library
of Australia.

ISBN 978 0 6485860 9 8

Printed by Ingram Spark
www.ingramspark.com

Designer, Illustrator and Typeset by
SUNRISE

After the law of love, the laws of electromagnetism rule the Universe.

CONTENTS

ACKNOWLEDGMENTS

The Australian National University has been kind over the years to provide access to various publications held in the library at the School of Physical Sciences, from the rare to the obscure, discussing Einstein's Unified Field Theory.

Whilst it can be hard to find publications discussing Einstein's final great theory, and even fewer written in a way that is simple to understand, some useful pages discussing the work have appeared in certain quality publications on Einstein's work and his biography from notable authors. These include people like Jeremy Bernstein, Robert Cwiklik, Philipp Frank, Dr Leopold Infeld, Dr Vaclav Hlavaty, Professor Tullio Levi-Civita, and Professor Stephen Hawking, to name a few. Their insights and contributions have been invaluable in understanding the key concept.

Finally, for those of you who will understand the concept and put it to use for the benefit of humanity, the acknowledgement also extends to you as the new caretakers of this knowledge. We hope the future you create for this world with this knowledge will give greater joy, stability, simplicity and a stronger sense of curiosity about life and the Universe to all people.

And, most importantly, what you do with this knowledge and its application will be through love.

CHAPTER 1

Introduction

As far as the laws of mathematics refer to reality, they are not certain; and as far as they are certain, they do not refer to reality.
—Albert Einstein[1]

THE TURN of the 20th century was a time of great upheaval in the world of physics.

The study of electromagnetism had only just been completed, resulting in the formulation of highly robust formulas for explaining the electromagnetic world. Indeed, the seemingly distinct electric and magnetic fields were finally unified into what physicists would call the "electromagnetic field" thanks to the brilliant work of James Clerk Maxwell (1831–1879). The result of all of this important work was the most complete and successful body of knowledge about electromagnetism ever devised in the world of physics—so complete[2],

---

1   Quote obtained from an address to the Prussian Academy of Sciences in Berlin delivered by Einstein on January 27, 1921.

2   Although no scientific subject is ever complete, to be called a law means that electromagnetism is a rare example of knowledge that has stood the test of time and is, therefore, considered to be complete.

1

Albert Einstein (1879–1955), Princeton, New Jersey, in 1947.
Photograph by Oren Jack Turner courtesy of the U.S. Library of Congress.

in fact, that many physicists would later describe it as a law rather than a theory.

Not so for other areas of physics, which had not quite reached their fundamental forms.

Among the most important areas requiring closer attention by scientists were new ideas about the structure of the atom; questions about why the speed of light as it propagated through the so called ether did not vary with respect to the motion of the Earth; and, of course, the great mystery of gravity and universal gravitation, which still beckoned for an answer.

Yet all of this was nothing compared with the new physics that was about to emerge. And it would disturb the greatest theoretical physicist of the 20th century, Dr. Albert Einstein (1879–1955).

It all began in 1905 when Einstein performed an experiment on the photoelectric phenomenon. In this experiment, he noticed tiny negatively charged atomic particles called *electrons* being ejected from the surface of a metal plate by something in light above a certain frequency[3]. One would have to say that this "something" was behaving like particles, albeit an invisible one.

Only problem was, physicists thought they had put the final nail in the coffin for the "particle-like" theory of light that Sir Isaac Newton first postulated in the 16th century. They were already opening the champagne bottle to celebrate the "wave-like" theory of light thanks to the work of Thomas Young and others when Einstein came along and spoiled the party. Now Einstein was not so sure if physicists should ignore the particle-like theory thanks to his discovery in the photoelectric phenomenon.

Determined to find out what these particles embedded in the light were, Einstein eventually arrived at the idea that light travels through space in packets of compressed electromagnetic energy called *quanta*. Each packet of energy is distributed over a certain volume of space. Never is the energy confined to an infinitely sharp point. For an infinitely sharp point to even vaguely be achieved, the frequency of light would have to be infinite. But, of course, Einstein knew this to be impossible. At any rate, this packet of energy somehow exerts a force

---

3   A plate composed of a different metal will hold onto the electrons differently due to the electrostatic forces that the protons in the nucleus of the atoms exert on the electrons. As a result, it requires different frequencies of the radiation to have enough force to eject the electrons.

on the electrons depending on the density of the energy (which is controlled by frequency) such that light behaves like a stream of particles.

Today, physicists feel quite at home in calling this particle-like effect of the light quanta by a more fancier name—a *photon*, for lack of a better word.

Despite suggesting this rather reasonable explanation in the face of the evidence (even if the name of this particle-like effect of the light quanta sounds more like it came out of a science fiction episode of *Star Trek* with no means of helping scientists to get closer to the truth of what this is), Einstein was not entirely satisfied. As he would discover in another experiment, the same force that light exerts also influences uncharged matter. Now the mystery had really deepened for Einstein. Unfortunately, Einstein did not have time to pursue a logical conclusion for the issue for at least another 10 years due to his fervent preoccupation with finding work to support a new family and later with solving the problem of gravity and universal gravitation. In the meantime, other physicists simply learned to accept Einstein's explanation and acknowledged the paradoxical nature of light. There was no expectation of any further work on the electromagnetic phenomenon.

Did this mean the scientists could finally celebrate now?

Not so fast, my dear Watson.

Just when things were starting to settle down, suddenly the atomic world began to make waves among some physicists with the revelation of just how incredibly lightweight a typical atomic particle really is. If you tried to observe, say, an electron using high-frequency light energy, you would discover how the light would disturb the electron enough to make it impossible to determine precisely where the electron went. Sure, you might have some idea of the initial speed of the electron at the precise moment of collision with the light. And you might have a good idea of its initial position because the volume of the light energy is small and creates a certain amount of force, which you can calculate. But what happens afterward is anyone's guess. Likewise, using a lower-frequency light energy might disturb the electron the least. Unfortunately, the energy is distributed over a larger volume, making it

Dr Werner Heisenberg (1901–1976), Professor of Theoretical Physics and Director of the Max Planck Institute for Physics in Göttingen. His contribution to quantum physics would influence Einstein's decision to pursue an alternative path to understanding the nature of light.

virtually impossible for physicists to measure the electron's exact initial position.

As a consequence of this latest discovery, a band of renegade "young gun" physicists headed by Dr. Werner Heisenberg (1901–1976) and his colleague Dr. Erwin Schrödinger (1887–1961), together with important contributions from Paul Dirac (1902–1984), thought the time was ripe in the early 1920s to create a new physics known as *quantum mechanics*.

Underpinning the foundations of quantum mechanics is the Heisenberg uncertainty principle. In the exact words of Heisenberg, he stated:

> "Die Unschärferelation: Je genauer wir den den Ort eines Teilchens kennen, desto weniger kennen wir seinen Impuls. Umgekehrt: Je mehr wir uber den Impuls kennen, desto weniger wissen wir über seinen Ort."[4]

Or for the more German-language-challenged among us, Heisenberg said:

> "The more precisely the position is determined, the less precisely the momentum is known in this instant, and vice versa."[5]

The consequence of the new physics was that no one could see what was happening to these pesky quantum particles. Forget the idea of drawing infinitely thin lines in space to describe the paths of every single quantum particle using the old and trusted equations of classical Newtonian science (including electromagnetism), and then observing these paths to confirm the mathematical results. Being so heavily reliant on the eyes to do all of the observing (and mathematics to get the precise paths), these physicists realized that it was simply not practical to do this anymore. So, why bother to understand the precise mechanism of what is going on in the universe of these tiny particles? Their idea was this: Just apply the new mathematical equation that Schrödinger developed for the quantum world to describe the probability of where one or more particles are likely to be positioned,

---

4   *This quote appears in Heisenberg's original paper published in Zeitschrift für Physik, 43 (1927).*
5   As quoted in "The Uncertainty Principle" at the American Institute of Physics.

as well as their speeds. That is about the most anyone can do. There is no need to bust your brains in determining precisely what an individual particle is actually doing in the real world. Indeed, if there is anything bizarre that scientists discover about the unseen quantum world through the mathematics of quantum mechanics, then we just have to accept it.

Now it seems that these physicists were heralding quantum mechanics as a major scientific achievement. In fact, so confident was this new breed of ambitious young physicists in their new science that they thought it was time to celebrate as if all scientific discovery had come to an end—at least for this area of physics. Except Einstein was, again, not joining the party.

Einstein saw a problem with the new physics. He observed how the new physics refused to delve further into the problem of why light moves uncharged matter. In fact, it would not even bother explaining the behavior of individual particles. Instead, physicists took the stance that if they could not see what was happening, what was the point of explaining the behavior of any individual particle? You might as well accept whatever mathematical solutions are derived from quantum mechanics and be done with it.

Einstein disagreed.

At the heart of the problem for Einstein was the manner in which these other physicists decided, as they still do to this day, to treat these particles like dice in a giant probability machine. And if we set up experiments to see what the particles would do next, they seem to know where to be positioned in space in a way that we do not expect under Newtonian physics. A classic example would have to be the modified Thomas Young experiment where two slits on a partition are opened to permit the passage of particles through them. However, instead of landing on a screen in front of the slits as two bright bands corresponding to the number of slits opened as physicists would expect if they were solid objects following Newtonian physics, the particles somehow know how to land in multiple bands right across the screen, not unlike the way in which the waves of light passing through the slits constructively interfere to create the same bright bands. Does this mean every particle has a mind of its own and knows how to behave in a wave-like manner when physicists set up certain conditions? Either that, or a mysterious God must somehow exist to influence the particles to behave in the way they do. Well, how else can we explain it? However, given the way in which quantum mechanics has been set up and formulated, we are forever forced to accept the bizarre results of the quantum world without ever explaining the reason why the particles behave the way they do (and hence find the simple "cause-and-effect" relation that Einstein believed should exist). Therefore, people like Einstein who want to find an explanation would be forced to consider exotic explanations, including the possibility that God may exist, that particles are alive and can decide what to do, or that new, exotic, "God-like" particles are influencing the results. The laws of probability that Schrödinger's equation governs are not concerned with the path that a single atomic particle takes. Rather, they do nothing more than give probabilities of how the particle is likely to behave and where it is likely to be positioned in space. In light of this, the laws of Newtonian physics could never be applied to the mysterious quantum world to calculate the precise path of any atomic particle and thus give the hapless scientists the chance to perhaps uncover the likely classical explanation for why certain quantum effects have been observed or predicted.

Einstein disagreed with the implications. There had to be a logical explanation.

To prove that he was right, Einstein decided he would solve the problem by developing another new physics. He would call it the *Unified Field Theory*. And it was here where he could tackle the ultimate physics issue on his mind: how light can move uncharged matter and eventually explain what is really going on in the quantum world.

Einstein would spend many valiant years and probably many sleepless nights on the new theory.

Finally, in 1929, Einstein published a completed 8-page paper titled "Zur einheitlichen Feldtheorie"[6] ("The Unified Field Theory") in a Prussian scientific journal[7].

On closer inspection, we can see that the aim of the paper was relatively straightforward: Einstein wanted to marry the electromagnetic and gravitational fields into a single unified field and see whether it

---

6   A more accurate English translation is "About the Unified Field Theory".
7   The completed theory appeared in *Sitzungsberichten der Preussischen Akademie der Wissenschaften* (translated as "Proceedings of the Prussian Academy of Sciences") in 1929.

could be used to explain all phenomena in the Universe[8], including the quantum world using simple "cause-and-effect" relations in a purely Newtonian sense.

As Einstein explained in his German-language book *Mein Weltbild* (with the quote republished in English in his book *Essays in Science* in 1934):

> "It would of course be a great step forward if we succeeded in combining the gravitational field and the electromagnetic field into a single structure. Only so could the era in theoretical physics inaugurated by Faraday and Clerk Maxwell be brought to a satisfactory close."[9]

Or, to put it simply, Einstein's scientific love affair with the laws of electromagnetism shows how the Unified Field Theory was really a bold attempt to create a unifying theory, just as Maxwell did for the electric and magnetic fields, except that here Einstein was extraordinarily ambitious. His dream was nothing less than to unify physics by combining the electromagnetic and gravitational fields to form a type of electrogravitational field. This was a truly stupendous effort in anyone's language.

After all of these years since the paper was published, a great question mark still hangs over his work: Did Einstein succeed in his attempt? Did Einstein somehow see the light at the end of the tunnel, so to speak? Well, if he did, other scientists were not privy to the answer. As a result, many scientists have decided that his work was a failure.

Whatever Einstein discovered, it was enough for him to stick by his final scientific masterpiece throughout the rest of his life. Indeed, on the night of his death in 1955, Einstein called his secretary to bring in his most recent calculations concerning the Unified Field Theory as if

---

8   The part that is visible to our eyes and instruments up to a distance of approximately 13.7 billion light years away is called the *visible universe* (or simply the *universe*). As scientists believe that the universe is experiencing an accelerating expansion, all mass-energy in the universe must be filling up a much larger space that we call the *Universe*. In Chapter 6, the Unified Field Theory suggests that this expansion is an optical illusion created when light naturally loses energy as it passes through the mass-energy of space. It has nothing to do with galaxies racing away from us. If this is true, the universe is effectively the Universe for good reasons as we shall discuss, and as such, we will use the term "Universe" throughout this book unless we specifically want to focus on the visible part of the Universe.

9   Einstein 1934, p.19; Childress 1990, p.11.

he were still clinging on to his great dream. If anyone knew whether he was successful, it would have been Einstein himself.

Perhaps Einstein was still grappling with the complexity of the mathematics in his final theory and wasn't getting anywhere? Or, maybe he was indicating to the world from his deathbed that he was on the right track but felt that the world was not ready to understand what he had discovered?

History suggests the latter situation. There is something Einstein had seen, and he was prepared to spend the rest of his life working on his final theory. As for any secrets he had found in his work, he decided not to tell a soul.

Whatever those secrets were that made him quietly confident, the question we must ask is, Was he successful in his attempt to unify physics?

To the mathematicians, there is no question: they all agree that the theory is a work of sheer genius. Nothing in the equations that Einstein wrote is incorrect or does not satisfy the conservation of energy principles.

As usual, let the covariant components of the energy tensor be denoted by $T_{\mu\nu}$. If influences of any origin are admitted, these quantities $T_{\mu\nu}$ are to be imagined broken up into two parts, one of which, $\tau_{\mu\nu}$, is purely electromagnetic, and the other, $\mathbf{T}_{\mu\nu}$, represents the remainder, if any. We therefore put

$$T_{\mu\nu} = \tau_{\mu\nu} + \mathbf{T}_{\mu\nu}, \quad \ldots \ldots \quad (32)$$

where τ is the well-known Maxwell tensor; further, for empty space $\mathbf{T}_{\mu\nu}$ is of course equal to zero.

As is well known, the Einstein equations (without the cosmological term) are

$$G_{\mu\nu} - \tfrac{1}{2} G g_{\mu\nu} = -\kappa T_{\mu\nu},$$

where the constant of proportionality κ may be expressed in terms of f, the gravitational constant, and c, the velocity of light $\left(\kappa = \dfrac{8\pi f}{c^4}\right)$.

The unified field equations are presented here in shorthand vector form to make them look less daunting for mathematicians. The gravitational field is mathematically represented on the left side of the bottom equation (some mathematicians further simplify this to the vector term $G_{\mu,\nu}$), and the electromagnetic field by the vector term $T_{\mu,\nu}$ shown on the right side of equation (32).

However, to the uninitiated, and to the physicists whose job it is to relate mathematics to reality through experimentation, the Unified Field

Theory appears to be a mathematical monstrosity of such epic proportions that no one can fully understand it, let alone relate the unified field equations to reality. This may explain why the *Encyclopedia Britannica* stated the following about the Unified Field Theory:

> "...Einstein and others attempted to construct a unified field theory in which electromagnetism and gravity would emerge as different aspects of a single fundamental field. They failed, and to this day gravity remains beyond attempts at a unified field theory."

In other words, a scientific view exists among Einstein's contemporaries that he was unsuccessful in his unifying attempts. Is this true?

In the meantime, physicists feel incredibly certain that quantum electrodynamics (QED) is the key to creating a "Theory of Everything". For example, in 1971, U.S. physicists Steven Weinberg and Sheldon Glashow, along with Pakistani physicist Abdus Salam, successfully combined the weak nuclear force governing the decay process of a neutron to a proton in the atomic nuclei and the electromagnetic force to form a single field theory known as the electroweak theory with the help of QED. Now, the physicists are working to combine the strong nuclear force for keeping protons together in the atomic nuclei with the electroweak force to create a grander unified field known as, quite logically enough, the grand unified theory. Later, with the help of the superstring theory, gravity may one day be incorporated into this grand unified theory to construct what is known as the *Theory of Everything.*

However, this book will present the thought experiments that made Einstein so certain that he had to develop his Unified Field Theory. We will also present the actual picture and concept that Einstein had figured out and considered essential to creating his Unified Field Theory. Indeed, from this concept, we can present the correct interpretation that one should give to his final great theory. Then, we will tackle the problem of, what is the gravitational field? Does it exist? And if so, what is it exactly? Finally, we will look at some of the expected implications of the concept when understanding the various mysteries of the universe, not least of which is how gravity and universal gravitation work from a new scientific standpoint. Consider it a kind of breath of fresh air for the stale, old physics presented by Sir Isaac Newton. But we will not stop there. We have an opportunity to tackle the size and age of the universe, understand the aging process for all living things, uncover a new explanation for what strong and weak nuclear forces are, and, finally, a look at the nature of God in religion. These are questions that have kept many people, including the physicists, preoccupied for many generations. It seems befitting that a book of this nature should include an attempt to view these mysteries in a brand new light.

Far from the idea that Einstein failed in his quest, strong indications now exist that Einstein was successful and simply kept this to himself. What the problem seems to be here is a lack of imagination from the rest of the scientific community to help them properly understand the Unified Field Theory and get to the very essence of what it is Einstein was trying to do. In fact, these physicists did not do the necessary work of pushing the idea to a logical conclusion to see the type of universe that we might be living in and the new explanations we can find for the various mysteries of the universe. If the physicists had looked more closely at Einstein's final scientific legacy, they may have realized the following:

1. Light obeys all of the classical laws of physics.
2. Light is gravity, and gravity is light (in fact, light is the unified field).
3. Light is the ubiquitous substance making up and permeating all matter and space, and it is the fundamental force of nature that moves and holds together all matter from the largest to the smallest scale.
4. Newtonian physics, relativity, and quantum mechanics (and later nuclear physics) would come together under the umbrella of electromagnetism, making electromagnetism the fundamental law of the Universe.
5. The ideas of determinism can be applied to all of physics, including the quantum world, no matter how difficult it is to observe on the smallest scale (it just requires physicists to use their imagination or perform relevant high computational

volume computer simulations to indirectly see what is really happening); and

6. To determine for those scientists, including Einstein, interested in the concept of God, whether the Universe came about as the result of some preordained plan (thus proving that God exists) or as a random event. As Einstein said: "I want to know how God created this world....I want to know His thoughts; the rest are details."[10]

Speaking of God, it is little wonder that an English theoretical physicist and the director of research at the Centre for Theoretical Cosmology within the University of Cambridge, Professor Stephen William Hawking (1942–2018), famously said that any theory to unify all of physics would be "the ultimate triumph of human reason—for then we should know the mind of God"[11].

Of course, this is not to say that Hawking believed or has given unequivocally supported the idea of the existence of God as we can see in the following quote from this brilliant English physicist:

"When people ask me if a god created the Universe, I tell them that the question itself makes no sense. Time didn't exist before the big bang, so there is no time for god to make the Universe in. It's like asking directions to the edge of the earth; The Earth is a sphere; it doesn't have an edge; so looking for it is a futile exercise. We are each free to believe what we want, and it's my view that the simplest explanation is; there is no god. No one created our Universe, and no one directs our fate. This leads me to a profound realization; There is probably no

Stephen Hawking

10   http://www.execpc.com/~shepler/einstein.html
11   Hawking 1988, p.193.

heaven, and no afterlife either. We have this one life to appreciate the grand design of the Universe, and for that I am extremely grateful."[12]

Rather, Hawking's statement about the "mind of God" merely showed that he was being open-minded on the issue in case Einstein should discover something that would link God to his work on the Unified Field Theory.

Michio Katu, a theoretical physicist at City College, City University of New York, echoed similar words when he said that anyone who succeeds with the Unified Field Theory should be able to show "an equation an inch long that would allow us to read the mind of God."[13]

The only thing is, did Einstein see the "mind of God" through his work? In other words, did the great man succeed in what he had set out to achieve?

There is only one way to find out, and that is to delve into Einstein's work. Nothing short of understanding what made him take on the mother of all mathematical monstrosities, how the unified field equations are structured and the basic way of interpreting them, and what was contained in those thought experiments that convinced Einstein to pursue his ambitious goal. But to make sure that we keep our heads above the mathematical waters, we will keep our explanations as simple as possible. Because the aim here is to ensure that the ideas can be related to reality.

We are not here to prove how clever we are with our mathematics, but rather how smart we are in seeing the reality of Einstein's work, which is the ultimate aim of the physicist. Mathematics can always come later when physicists apply the idea from Einstein and create simpler pictures of the universe. And from there, the sizes of the forces in question will hopefully match those in reality. Let us simply focus on the main picture in Einstein's mind—the one that led him to develop his Unified Field Theory.

With this in mind, this book explains how far we have come in creating a unifying theory for physics thanks to Einstein's final great scientific masterpiece.

---

12   https://www.goodreads.com/quotes/551152-when-people-ask-me-if-a-god-created-the-universe.
      Quote appeared in the documentary *Curiosity* televised on the Discovery Channel.
13   http://whatis.techtarget.com/definition/unified-field-theory-or-Theory-of-Everything-TOE.

CHAPTER 2

The Development of the Unified Field Theory

It is a miracle that curiosity survives formal education.
The only thing that interferes with my learning is my education.
—Albert Einstein[1]

ALBERT EINSTEIN was born in the town of Ulm, Germany, on March 14, 1879, to Hermann Einstein (1847–1902) and Pauline Koch (1858–1920). On June 21, 1880, Albert's family moved to Munich, where his father and uncle founded *Elektrotechnische Fabrik J. Einstein & Cie*, a small shop selling manufactured electrical equipment powered by direct current.

---

1   The original source is unknown, but the general consensus is that the quote did come from Einstein. Another similar quote attributed to Einstein was known to have been made on March 12, 1949. When asked for his thoughts on the educational system as he understood it, Einstein said (and was later published in the *New York Times* the following day), "It is nothing short of a miracle that modern methods of instruction have not yet entirely strangled the holy curiosity of inquiry."

Albert Einstein at age 3 years.

The family, including Albert's bright young sister Maria "Maja" Einstein (1881–1951), lived in a relatively modest house not far from the shop.

Albert—The Curious Boy

It did not take long for the family to notice something special in young Albert. By the age of 3 years, Albert was already displaying the characteristic hallmarks of a great young scientist, particularly his pronounced curiosity about his environment. Whether his close-knit and loving family strongly encouraged this or this was a natural gift, by the age of four, the young boy was already insatiably exploring his father's shop.

As any young boy would know, a shop filled with interesting electromagnetic and mechanical gadgets is like a visit to the home of Father Christmas at the North Pole. Here was a veritable treasure trove of what seemed to be an endless supply of gadgets to entertain and capture the imagination of any young boy. Of course, as with any young boy curious about an intricate toy, there would always be the temptation to break it apart to see how it worked. Knowing this was a possibility, Einstein's father had other ideas for his young and precocious son. At the same time, he may have wanted his son to take over his shop one day—a natural aspiration. Before this could come about, young Albert needed to know something about how the devices worked.

So his father played a special game with Albert. It would be a game to see how well his son could figure out the inner workings of the devices in the shop without looking inside of them straight away. Albert was the perfect candidate for this game thanks to his curiosity and strong visualization skills. It was clear right from the outset that Albert wanted to know how things worked and to show his father that he knew the answer, and he was prepared to do it using his mind. As he himself put it:

> "From an early age I loved to look at machines, and understand how they work."

And he did exactly that. Instead of immediately looking inside and tinkering with things, Albert sought to understand and visualize the problem first, then later confirm his thoughts by using his eyes to look inside of the various machines.

With practice, Albert became very good at this game of determining how things worked. At the same time, the young boy began to develop a trust in the theory of how something worked, rather than taking the hands-on approach of getting inside and looking at things. He would start with what he knew, and with his imagination he would create a picture in his mind of how something had achieved its seemingly wondrous and extraordinary behavior. Sometimes he asked his father for a clue if he felt unsure. When Albert did figure out how a gadget worked, he would receive praise from his father and uncle, then later from the rest of his family.

Eventually, a point came in Albert's life when his father was impressed with his son's ability to quickly figure things out on his own without any clues. Then, one day, when five-year-old Albert was sick in bed, his father gave him something to test his finely tuned abilities in finding an explanation. That something turned out to be a magnetic compass.

Curiosity and the Compass

The spark that started Albert on a life-long journey to become a great theoretical physicist happened in 1884, all because of the compass his father had given him.

As history would tell us, Albert was left alone in bed to ponder over the strange instrument. He was thoroughly captivated by the compass. This was something he had not seen before. At first, he thought it was some kind of special clock with a long metal arm attached to a central shaft. But as soon as he turned the wooden box and saw how the metal pointer would swivel on the tip of a fine central pin, he noticed a strange behavior: the arm was determined to point in a particular direction.

Albert must have wondered what caused this piece of metal to move in such a determined way. Perhaps a tiny motor was sitting inside of the box? Nope. When his father gave him permission to look inside, he could see nothing. How could that be?

Here lay a deep problem. Were ghosts holding the needle in place? Surely not. Einstein didn't believe in ghosts. He knew there had to be a logical and scientific explanation for this behavior. As Einstein recalled:

> "I experienced a miracle when my father showed me a compass. I trembled and grew cold. There had to be something behind objects that lay deeply hidden."

As he couldn't figure out what was causing the needle to point in a constant direction, he pestered his father one more time for an explanation. But this time, his father wasn't too sure either. Although a brilliant mathematician named James Clerk Maxwell determined the laws of electromagnetism, his father was not into this type of advanced work. He was more interested in the practical aspects of fixing and selling electrical devices. The intricate details of the theory of electromagnetism—including the presence of the magnetic field surrounding the Earth and its influence on magnetized metals, such as iron—were not his forté. At last, Einstein had found something that his father did not understand. Now could he really impress his father by figuring out how the compass worked?

From that moment, young Einstein knew exactly what he wanted to do. He would embark on a quest to understand how the needle in his compass moved in such a determined way, and see where this work would lead him.

Einstein's Great Lifelong Mentors

Einstein's persistent boyhood interest in electricity and magnetism eventually led him to study the work of two great British scientists. It was Michael Faraday (1791–1867) who had masterfully carried out many of the important practical experiments in electricity and magnetism[2]. But young Einstein was more fascinated with the actual theories of electricity and magnetism, which helped to explain the results of these experiments. This was James Clerk Maxwell's unquestionable expertise. So, when Maxwell completed the theory of electromagnetism in 1864, specifically with regard to four elegant mathematical equations (see Appendix B for more details),

---

2   Hans Christian Ørsted was the first person to see a link between electricity and magnetism. In 1820, Ørsted observed how electric currents exert forces on magnets.

James Clerk Maxwell

$$\nabla \times \mathbf{H} = \mathbf{J} + \frac{\partial \mathbf{D}}{\partial t} \quad \textit{(Ampere's Law)}$$

$$\nabla \times \mathbf{E} = - \frac{\partial \mathbf{B}}{\partial t} \quad \textit{(Faraday's Law)}$$

$$\nabla . \mathbf{D} = \rho \quad \textit{(Gauss' Law for Electric Field)}$$

$$\nabla . \mathbf{B} = 0 \quad \textit{(Gauss' Law for Magnetic Field / Non-existence of monopole)}$$

young Einstein immersed himself in Maxwell's work[3].

Michael Faraday.

From this work, Einstein quickly realized that the compass behaved as it did because the Earth generated a magnetic field and was influencing the magnetized iron arm to point in a particular direction.

Einstein didn't stop there, though. Further detailed reading into Maxwell's work also revealed another amazing fact: how quickly the

---

3   Readers are not required to memorize these equations. You will not be tested on them, or anything like that. We just present the equations to show that they do exist.

mysterious electromagnetic field in Maxwell's equations can travel. It can move at phenomenal speeds. Today we understand the speed to be 300,000 km/s. Back then, this seemed so fast that the average person on the street could not begin to imagine it (and neither could many scientists at the time).

Not so for Einstein. The young boy with so much curiosity had the power to visualize and imagine, so much so that he began to picture in his fertile young mind what it would be like to travel on the crest of a light wave.

The Special Theory of Relativity Comes of Age

Years would pass as Einstein fine-tuned his mathematical abilities and applied extensive visualization skills through the careful use of his imagination to create what we now call "thought experiments" on the problem of high-speed physics. Einstein's persistence and vigilant research eventually paid off when he worked out a brilliant and revolutionary new theory to explain what happens when something travels at nearly the speed of light. This theory was revolutionary because it broke the traditional concepts of time, length, and mass for a moving object.

Before Einstein tackled the problem, many scientists thought that any measurement of:

1. the length along the direction of motion for a moving object was the same at all speeds;
2. the mass of a moving object was constant, and could be made to travel at infinite speeds; and
3. time was totally independent of the motion of the mass and that the time flowed on uniformly.

This was the accepted scientific view by the end of the 19th century thanks to the work of Sir Isaac Newton (1642–1727) and his laws of motion developed nearly 400 years earlier.

In slightly more mathematical language[4], this would be equivalent to saying,

$$t_1 = t_o$$
$$m_1 = m_o$$
$$L_1 = L_o$$

for any object passing through space at virtually any speed. In other words, if an object at rest relative to an observer has a mass of m_o, if it has a length of L_o, and if the time measured for anything attached to the object is t_o, the same observer taking another set of measurements of the same object at a faster speed would obtain identical values for m_1, L_1, and t_1.

Now Einstein would challenge all of these ideas.

In fact, what Einstein discovered after developing his theory is that as an object's velocity, v, approaches c=300,000 km/s (kilometers per second) or 300,000,000 m/s (meters per second), known as the *speed of light*, three things happen:

1. Time dilation

The measurement of time t_o on all artificial and biological clocks attached to the moving object, as seen by an outside observer who thinks he is at rest with respect to the moving object, will appear to slow down to t_1 as given by the following equation:

$$t_1 = \frac{t_o}{\sqrt{1 - \frac{v^2}{c^2}}}$$

---

4   Don't panic! For most readers, you can safely ignore these equations. For those who are curious, imagine that the letters of the equations you see on this page and anywhere else in this book are numbers. Even L_o is a number too. If you would like, you can label L_o as either L, b_o or K. It does not really matter. Just imagine these letters as numbers. Now some of these numbers may represent certain things in the real world. They could be related to mass, time, length or velocity. These can be measured in seconds, minutes, hours or doggy years for time if you would like. Mass can be measured in grams, kilograms, tons or whatever. The same with length and velocity. So you can substitute, say, the letter v for velocity with a number like 50 km/h. But if you do, make sure that you keep the units of measurement consistent. Indeed, scientists are particularly fond of keeping the units of measurement the same. So do not be surprised if scientists prefer to measure velocity in terms of m/s or m.s^{-1} (different ways of writing meters per second). Likewise, mass is usually measured in kilograms and time in seconds. In this way, anyone can reproduce the answer that comes out of an equation, and the answer can be related more closely to reality.

2. Mass increases

The measurement of mass m_o of the object when it was formally at rest will appear to increase to m_1 (i.e., the object will gain mass) when it moves as given by the following equation:

$$m_1 = \frac{m_o}{\sqrt{1 - \frac{v^2}{c^2}}}$$

3. Length decreases

The measurement of length L_o of the object along the direction of motion will appear to diminish to L_1 when it is moving as given by the following equation:

$$L_1 = L_o \sqrt{1 - \frac{v^2}{c^2}}$$

according to an outside observer making measurements of the moving object.

To give an example, suppose you sat in a spaceship moving at, say, 90 percent of the speed of light. That is 90 percent of 300,000,000 m/s, which is 270,000,000 m/s. Because you are moving at the same speed and in the same direction as the spaceship, your measurements of the mass and length of the spaceship, as well as your perception of the time onboard and how it flows based on a clock you can look at will appear to be the same, as if the spaceship were not moving at all. Indeed, if you had no windows to the outside world, you could assume that the spaceship was not moving.

However, someone sitting on a planet using a telescope and observing you traveling through space will measure mass, length, and time as follows:

$$m_1 = \frac{m_o}{\sqrt{1 - \frac{v^2}{c^2}}} = \frac{100 \, \text{kg}}{\sqrt{1 - \frac{(270,000,000 \, \text{m/s})^2}{(300,000,000 \, \text{m/s})^2}}}$$

$$= \frac{100 \, \text{kg}}{0.43588989} = 229.4 \, \text{kg}$$

This means anything that weighs 100 kilograms at rest inside or attached to the spaceship will appear to weigh about 229 kilograms at 90 percent of the speed of light as measured by the outside observer looking at the moving mass.

As for the length along the direction of motion of anything attached to the spaceship, including the spaceship itself, this will appear to contract according to an outside observer who is taking the measurement while at rest. If, for example, the length of the spaceship measured from the back to the front is 100 meters long, traveling at 90 percent of the speed of light causes the outside observer to measure a contraction in the length to 43.5 meters.

$$L_1 = L_o \sqrt{1 - \frac{v^2}{c^2}} = 100\text{ m}\sqrt{1 - \frac{(270,000,000\text{ m/s})^2}{(300,000,000\text{ m/s})^2}}$$
$$= 100\text{ m x }0.43588989 = 43.5\text{ m}$$

As for time, a passage of 100 seconds at rest would be equivalent to 229.4 seconds for the person observing you inside of the spaceship.

$$t_1 = \frac{t_o}{\sqrt{1 - \frac{v^2}{c^2}}} = \frac{100\text{ s}}{\sqrt{1 - \frac{(270,000,000\text{ m/s})^2}{(300,000,000\text{ m/s})^2}}}$$
$$= \frac{100\text{ s}}{0.43588989} = 229.4\text{ s}$$

As intriguing as all of this may seem, some readers might wonder why the outside observer sees these things occur. Indeed, what does this mean for a pilot sitting in the moving spacecraft when we watch him? Will the extra weight crush him, will the length of contraction squeeze him, and will he appear to suffer a slow speech impediment when he speaks? Don't worry. Everything is fine for the pilot. If you could go there instantly to sit right next to him, you would see that he is fine. When we talk about time dilation, length contraction, and mass increase, we are referring to how the mass-energy of space that is surrounding the moving object affects the light (or radiation) that the object emits. We need light to observe and make measurements. The movement of the spacecraft simply distorts the light it emits such that it creates these strange effects.

What is actually happening is that part of the mass-energy[5] of space (made mostly of electromagnetic radiation) is displaced as it becomes attracted to the moving spaceship, especially around the front end, when moving linearly through space. It is not unlike the way in which a plough moves the dirt or snow in front of it. But at the same time, the moving spaceship is attracting this mass-energy to itself in a gravitational sense. As this mass-energy becomes concentrated, this extra mass-energy collides to a higher degree with the atoms comprising the moving object. This causes the atoms to vibrate more vigorously and to emit more mass-energy to the surroundings by way of radiation, which, in turn, increases the mass-energy and with it amplifies (or increases) the object's gravitational field. It is this increase in the gravitational field stemming from the concentrated mass-energy of space that gives the impression of a more massive object to the outside observer.

In addition, the concentrated mass-energy is a transparent region that can bend light from distant objects located in space along the direction of motion, thus creating what is effectively an optical illusion, where the objects in front of the spacecraft appear to be closer than they really are. It is like having a giant magnifying glass sitting in front of the spacecraft. As a result, the measurements of the lengths to the objects lying directly in front of the spacecraft will appear to be "contracted" (or be shorter). But in reality, the objects will still have the same distances. Remember, the true distance to any destination in space is always the same whether you are at rest or moving at a high speed. Never does the spacecraft and everything inside of it get squashed along the direction of motion[6], nor do the stars and planets along the direction of motion become magically jolted from their positions and

---

5   Physicists call it *mass-energy* because when the energy collides with ordinary solid matter, it moves the matter. It is as if the energy has mass and, in a sense, it is impossible to tell the difference with ordinary matter colliding. So, you can call it mass or energy, or you can see it as ordinary matter, whichever you prefer.

       As for what the mass-energy of space is, we know it contains electromagnetic energy called radiation, and it contains gravitational energy. Other forms of energy could exist, but what mass-energy is exactly, including the mystery of gravitational energy, was not elucidated in Einstein's Special Theory of Relativity or the General Theory of Relativity. It would require the Unified Field Theory to explain the mystery.

6   Although you will feel greater inertial forces on your body and spaceship during acceleration, in Chapter 3 we will learn about a way in which to reduce the inertial forces of anything sitting inside of the spaceship during acceleration, as well as gravity itself.

brought closer together via the moving spaceship. Everything in the Universe and inside of the spacecraft is located in the same place. It is just that the displacement of the mass-energy bends light from the objects directly ahead of the spacecraft to make them look as though they are much closer. But this is not true. The perceived distance is not the true distance. Rather, the mass-energy of space is bending the light to make things look closer to the observer inside of the spacecraft. At the same time, the displacement of this mass-energy helps the rest of the mass-energy of space to pull[7] the spacecraft to a higher speed, beyond the speed of light.[8] What differs is the speed of the spacecraft. To the outside observer looking at the spacecraft, it will appear to be traveling close to the speed of light. From this perspective, the spacecraft would take years to travel to, say, a star. However, for the person sitting in the spacecraft, the distance will appear to be shorter. And yet somehow, the spacecraft will reach its destination much more quickly, as if the distance was shorter. Well, that is because the spacecraft is being pulled to a higher speed. Think of the mass-energy of space as a stretched rubber band. Any reduction in the energy density from another region of space due to the displacement of this energy to concentrate itself around the spacecraft is effectively "stretching a rubber band". The rest of the mass-energy in space will quickly try to balance this by "pulling" on the higher mass-energy density around the spacecraft and the spacecraft itself by the gravitational field of this mass-energy of space in an attempt to fill the lower density region of space. But as soon as it does, the moving spacecraft will continue to concentrate this mass-energy from another region. Therefore, the spacecraft will keep getting pushed/pulled to a higher speed. This is how the spacecraft will reach its destination more quickly.

As for time, the dilation effect that an outside observer sees is due to the radiation emerging from the compressed mass-energy in front of or behind the spaceship. As it emerges, the wavelength of the radiation is suddenly much longer (or "redshifted") than when it was initially

---

7   According to the Unified Field Theory, this pulling effect that is described as the gravitational force of the mass-energy of space is really an electromagnetic pushing force from the radiation coming from behind the spacecraft.

8   The spacecraft can travel faster than 300,000km/s because the energy density of space has gone down. And within this region, both light and solid matter can travel faster, but never can solid matter exceed the speed of light. Whatever the speed of light in a lower energy density region of space is, the spacecraft must remain below this speed.

emitted from the spaceship. Part of this is due to the energy loss that occurs in the light as it moves through the compressed mass-energy, and this is translated as a stretching of the light (or a lowering of the frequency). The other part is simply because as the radiation moves away from the spacecraft, the density of the compressed mass-energy decreases, thus causing the wavelength to naturally increase. By the time the radiation emerges from this compressed mass-energy and moves through the normal density of mass-energy for the rest of space, the radiation will already have a much longer wavelength than it did when the spacecraft emitted it. Consequently, the outside observer, upon receiving this radiation, must wait longer than usual for all of the information in the light to arrive to see what is happening inside of the spaceship. If he does not, he will see you inside of the spaceship moving as if you were in slow motion.

However, from the perspective of being inside of the spaceship looking back at home or even at the destination planet, light that is sent from the outside and that passes through the higher density mass-energy that the moving spaceship creates along the direction of motion will have its wavelength shortened. Any energy loss in the radiation traveling through this compressed mass-energy is insufficient to bring it back to the normal frequency that the outside observer sends. Therefore, if you, while sitting inside of the spaceship, were to pick this up with a receiver and translate it into a moving image on a screen to see what is happening, it would be like watching a movie being played at high speed.

To highlight the radical nature of Einstein's idea, scientists use the famous "twin paradox" experiment, in which one person travels at nearly the speed of light and returns home. Meeting his twin, he discovers him to be much older than he is.

Einstein's paper containing these remarkable scientific results was sent to the leading German scientific journal, *Annalen der Physik*, under the title "On the Electrodynamics of Moving Bodies" on June 10, 1905, and was published on September 28, 1905.

Today, we call this the first stage of Einstein's Special Theory of Relativity.

The Famous E=mc² Equation

The second and final part of the theory came when Einstein published on September 27, 1905, another scientific paper discussing the total energy of an object. In this paper, Einstein showed the concept of the apparent total energy E of an object of mass m, and momentum p when the object is moving, through the following equation (see Appendix A for its derivation),

$$E = \sqrt{(pc)^2 + (mc^2)^2}$$

where c is the speed of light. In other words, we can use this formula to measure the amount of energy concentrated around a moving object, as well as the energy contained in the mass when it is at rest.

The equation is remarkably simple. Just to hit home the simplicity of the equation, the implication of this important formula for science can be better understood when we focus on a non-moving object. If we set momentum p (a measure of the object's speed) to zero (i.e., $p = mv = 0$ because velocity v is zero), which means the object is at rest, the total energy E of a non-moving object is given by the world-famous equation:

$$E = mc^2$$

This is a remarkably simple equation. So simple, in fact, that many physicists were initially shocked by the result and tried all they could to dismiss it. Obviously, it would not be for the mathematics of it. It was correct throughout Einstein's entire scientific paper. Rather, it was only because they could not clearly visualize the problem to see how true the equation was.

So what's the equation telling us?

It is essentially this: the equation reveals that a tremendous amount of energy is held:

- inside of the chemical bonds holding atoms together;
- in the so called nuclear forces holding together the subatomic particles consisting of the electrons, protons, and neutrons inside of the atom; and

Atomic bomb test, Bikini Island, 1946.

- in the composition of these subatomic particles (i.e., matter itself),

more so than anyone had ever imagined.

But this implication was not entirely clear among many of the more pragmatic physicists with a panache for physical experimentation at the time. It was only after the development of the atomic bomb in the late 1940s—in which a fraction of this massive amount of energy was released through a reordering of the protons and neutrons inside of the nucleus of an atom—did all of these scientists finally accept the reality of the equation. It was almost as though these scientists needed to see the equation played out in reality before they could ever take the idea seriously.

To add to the difficulty, the recognition of Einstein's Special Theory of Relativity was slow to come, mainly because he held no professorship or other reputable position of scientific standing. Einstein was merely a patent inspector at the patent office in Bern, Switzerland. So it was possible that scientists did not expect to see an independent person publishing original ideas in a scientific journal. Seriously, how could one man do this without help from other scientists? And how could his ideas be right?

But as more and more scientists came to understand and appreciate what he had achieved[9], he was eventually offered a place as an assistant professor at the University of Zurich, Switzerland, in 1909. Later, he would become a full professor at the University of Prague in 1911. And by 1913, he had accepted a directorship at the Institute of Physics in the Kaiser Wilhelm Institute in Berlin, Germany, which freed him of his teaching responsibilities.

Scientists may have initially ignored Einstein's efforts in favor of more team-based scientific work among familiar intellectuals in the world of physics, but nothing was going to stop Einstein from doing what he loved. Indeed, he had another pressing issue in the world of physics to solve.

---

9   German theoretical physicist Max Planck (1858-1947) helped to influence others in the scientific fraternity after he had noticed Einstein's paper, Without his help, Einstein could have continued to work as a patent inspector for many more years as if nothing unusual had happened. Part of the problem for Einstein was that no experiment existed to confirm his theory. The mathematics might have been right, but the ideas were considered to be too radical to accept immediately without some type of an experiment to support those ideas.

The General Theory of Relativity Comes of Age

Before the period when he was being offered professorships in various universities throughout Europe and abroad, Einstein had already turned his attention to the mystery of gravity as first revealed by the great 17th-century British mathematician, physicist and astronomer Sir Isaac Newton.

"Why gravity?" you might ask. Well, it is because there exists a rather unmistakable relationship between the mass of an object and its mysterious gravitational field—a fact that Newton first noticed. Basically, the more massive the object, the stronger the gravitational field it creates, even if the precise mechanism of how this mysterious field is created or what it is were not known. However, Einstein noticed something else.

Sir Isaac Newton

Albert Einstein in the Patent Office

In his Special Theory of Relativity, Einstein discovered how the mass of an object increases with speed. So, naturally, he would have thought to himself, "How would this affect the gravitational field? Surely it must increase in strength according to the classical law of universal gravitation as proposed by Newton". Yet the gravitational field for the same object at rest is apparently not as strong. Something about the way the object moves somehow controls the gravitational field. But why?

And more importantly, Einstein must have wondered, "What is the gravitational field?"

Or to put it another way, why does uncharged matter appear to be mutually attracted or "clump together" when brought in close proximity?

In November 1907, after much perseverance and determination, Einstein uncovered another important clue (apart from speed controlling the gravitational field) he needed to help formulate his new theory of gravity. Einstein explained his breakthrough regarding the problem as follows:

> "The breakthrough came suddenly one day. I was sitting on a chair in my patent office in Bern. Suddenly a thought struck me: If a man falls freely [i.e., in his own accelerated reference frame], he would not feel his weight [i.e., no gravitational field]. I was taken aback. This simple thought experiment made a deep impression on me. This led to the theory of gravity...."

Thanks to this simple thought experiment, Einstein made the next decision to see acceleration and the gravitational field as physically equivalent. As he stated in the 1907 edition of *Yearbook of Radioactivity and Electronics*:

> "We shall therefore assume the complete physical equivalence of a gravitational field and a corresponding acceleration of the reference system."[10]

In 1908, with the help of his close friend and former schoolmate who became a mathematician, Dr. Marcel Grossman (1878–1936),

---

10   From Einstein's article in Johannes Stark's *Yearbook of Radioactivity and Electronics* (1907).

Einstein summed up the results of his previous Special Theory on Relativity in what mathematicians would call the geometric "four-dimensional world view" form. He then extended the theory to take into account the reference frame with acceleration (to help encapsulate his new idea). And by November 25, 1915, Einstein had the complete results published in the same journal as his previous "special case" theory.

Without ignoring the efforts of other people in helping Einstein to achieve his work, Russian-German mathematician Dr. Hermann Minkowski (1864–1909) is credited with laying the mathematical foundation for Einstein's geometric approach to the theory of gravity and relativity.

Despite coming up with a solution, albeit a highly mathematical one using differential calculus and non-Euclidean geometry (shortened to *differential geometry*[11]), Einstein was already in trouble. He realized he was

---

11   Einstein argued that the laws of physics must be the same no matter what coordinate system is used. To support this view, he initially used differential geometry (of the non-Euclidean type) to develop his gravitational field equations of the General Theory of Relativity. Later, Einstein turned to tensor analysis —a newly-invented mathematical technique that Professor Tullio Levi-Civita and Gregorio Ricci-Curbastro conceived. The reason for making this switch can be seen in a letter he wrote to Levi-Civita:

> "I admire the elegance of your method of computation; it must be nice to ride through these fields upon the horse of true mathematics while the like of us have to make our way laboriously on foot."

Tensor analysis (or absolute differential calculus) and differential geometry are both aimed at providing a geometric description of the Universe in a way that is independent of coordinate systems. The latter avoids the use of any coordinate system (such as Cartesian or polar coordinates) altogether, whereas the former starts off with a coordinate system but never actually specifies what this is and it never relies on any special features of the coordinate system. Both techniques offer their own strengths and weaknesses, depending on the problem at hand to solve.

In the case of tensor analysis, the weakness can be explained by referring to Elie Cartan, when he stated that people should "...avoid very formal computations in which an orgy of tensor indices hides a geometric picture which is often very simple" (as published in *Riemannian Geometry in an Orthogonal Frame*).

On the other hand, relying too much on a completely independent coordinate system in differential geometry has its own problems, as Herman Weyl stated

> "In trying to avoid continual reference to the components, we are obliged to adopt an endless profusion of names and symbols in addition to an intricate set of rules for carrying out calculations, so that the balance of advantage is considerably on the negative side. An emphatic protest must be entered against these orgies of formalism which are threatening the peace of even the technical scientist."

Therefore, it is often recommended that anyone who goes into mathematics to learn these sorts of tools of the trade should learn both techniques and to know when to use one or the other when finding solutions to a problem. A combination of the geometric picture of differential calculus and the elegant analytical technique that tensor analysis offered is often the best approach.

Even after both techniques are mastered, the problem of relating mathematical solutions to the real world can still pose a significant problem to those whose job it is to relate mathematics to reality (i.e., the theoretical physicist). Mathematicians are only there to provide new mathematical techniques to make it easier to solve mathematical equations, not necessarily to relate the solutions to reality.

At any rate, the choice that Einstein made was to move from differential calculus to tensor analysis, as he felt that this was the best way in which to present his final published version of the gravitational field equations of the General Theory of Relativity, and later the unified field equations of

departing from his familiar and much-loved childhood skills of visualizing problems, creating simple thought experiments, and seeing the "picture" before applying basic mathematical equations to support it. At first, Einstein's geometric approach to general relativity did not rest well with him due to its sheer complexity[12]. But his colleagues convinced him to accept the mathematics behind his General Theory of Relativity. Once he felt comfortable with the mathematics and saw how he could combine his ideas in a unified way, he eventually accepted the mathematics as well. As American physicist Dr. Jeremy Bernstein said:

> "...Einstein was still allergic to pure mathematics, and for several years he was not particularly enthusiastic about Minkowski's "four-dimensional world view". It was only when he found the final formulation of his theory of gravitation, a sweeping generalization of Minkowski's work, that he fully appreciated its formal power."

So, what does the formal power and elegance of Einstein's mathematical formulation of gravity, universal gravitation, and relativity reveal to the world, preferably using the simplest language possible? Using the unmistakably clear and unambiguous language of the mathematician (cough!), Einstein's General Theory of Relativity—also known as the Geometric Theory of Gravitation—is described as an "apparent" force resulting from the bending of the four-dimensional space-time world via "accelerating" matter.

Still confused? Never mind. At least the mathematicians have listened and have created what they believe to be a much simpler explanation.

Imagine that you could simplify the Universe into two dimensions. We will use an infinitely thin and unbreakable piece of rubber fabric stretched out in all directions to represent the supposedly empty space throughout the Universe. This sheet of rubber also represents

---

the Unified Field Theory.

12   This does not imply that Einstein was bad at mathematics. Far from it. The aim of all theoretical physicists is to find the simplest solutions and to apply the simplest mathematical tools to achieve their ends. To say that a mathematical technique is complicated need not carry with it the connotation that Einstein never understood the mathematics or did not attempt to use the mathematics. Einstein always preferred to find mathematical tools that simplified the process, made sense in his own mind based on how the tools worked, and reduced the amount of mathematical symbolism involved in getting to a solution.

Einstein's concept of space-time, which is essentially the mass-energy of space. Remember, the space above the Earth's atmosphere is not a perfect vacuum. It has energy, mostly by way of electromagnetic radiation together with smaller numbers of other free moving particles, namely electrons and atoms. Other forms of "exotic" energy could exist, but the predominant energy in space is electromagnetic and gravitational, with the latter being the one that Einstein was trying to understand.

Now, if the fabric exists and remains flat, scientists interpret this as a gravitational field whose strength is constant throughout space. In other words, the mass-energy in space has a gravitational field, suggesting that either it creates its own gravitational field, or is carrying this gravitational energy from certain sources—mainly from ordinary uncharged matter. Of course, if there is absolutely no rubber sheet to be seen anywhere, not only does this mean that no mass-energy exists, but also it implies that no gravitational field exists either. However, for the purposes of explaining the concept, let's imagine we have a gravitational field (and, therefore, mass-energy is present and permeating all of space).

Next, we place an object of a certain mass[13] on this sheet. It will have to be in two dimensions as well. So let us imagine we have a flat coin sitting on the sheet.

Now if the countless numbers of atoms composing the infinitely flat object as well as the object itself do not vibrate, spin, or cause any other form of acceleration whatsoever, the two-dimensional mathematical fabric of space-time as represented by this rubber sheet will remain absolutely flat. Then, we have a situation where the object is not generating a gravitational field of its own to influence the mass-energy of space, nor is the gravitational field of space (via the rubber fabric) influencing the object. This means the object would float effortlessly in space.

However, should the object itself accelerate by either vibrating or spinning, and/or should the atoms in the object perform their own form of acceleration, it will generate a gravitational field of its own. When this gravitational field is *added* to the already pervasive gravitational field of space-time (or mass-energy) that has constant

---

13   Because ordinary solid matter contains mass, it, too, carries energy. It is just another form of mass-energy, albeit with a higher energy (or mass) density.

strength, the two-dimensional flat fabric of space-time bends in the presence of the mass. This "bending effect" provides an indication of the added strength (or reduction) taking place within the gravitational field.

How an object accelerates, as well as its direction, is just as important as how much accelerating mass is contained in the object. Acceleration can either increase or decrease the strength of the gravitational field[14].

Imagine yourself standing in an elevator. Should the elevator accelerate toward the Earth, you will notice that the weight of your body has decreased as if the mass has dropped or if the elevator is an anti-gravity device in reducing the gravitational field passing through the body and pushing you down to the ground. If you didn't know you were accelerating, you might be tempted to think the latter. However, should the elevator accelerate toward the sky, your body will feel as though it has more mass, and thus, you will feel heavier in a moving elevator. Otherwise, you might think that the elevator has amplified the strength of the gravitational field in and around your body and the elevator itself.

As you can see, whether an elevator increases or decreases the gravitational field strength depends on the direction and how fast the acceleration actually is. Either way, you will notice how much or how little gravitational force you will experience on your body when the gravitational fields of the Earth, the elevator, and your body are combined.

In mathematical terms, this change in the strength of the gravitational field is essentially described as a "bending of the curvature of space-time".

Thus, the more this fabric bends (i.e., the greater the acceleration is and/or the greater the amount of accelerating mass added to the object is) the stronger (or potentially the weaker) is the gravitational field is in the region where the object is located. This means that the curvature of space-time will influence any other accelerating object moving nearby,

---

14   It is important for the reader not to interpret this as saying that acceleration creates gravity as if this is the source of the gravitational field. All acceleration does is affect the strength of the gravitational field; never does it create the field. How acceleration actually affects the gravitational field is not fully explained in the General Theory of Relativity other than it changes the density of mass-energy in space. Until the mysterious mass-energy of space for controlling the gravitational field is explained, all that Einstein could say by 1915 was that acceleration is helping ordinary matter to experience the effects of the gravitational field.

as it will behave not unlike a ball rolling around the valleys of space-time.

If you are able to keep up with this explanation so far, you are a very special person indeed. But wait! There is more. What you have just read is the easy part. Next, try to imagine this same fabric in three dimensions representing space (i.e., our real Universe). Then, add a fourth dimension called time because the gravitational field affects time as well, and you will have some idea of how the mathematics of general relativity work.

Yes, we thought we would have confused the average reader with this mathematical interpretation.

Let us approach the concept in a different way.

A far less abstract description would be to say that wherever a *mathematical bending* of the four-dimensional space-time world occurs, there is, in fact, an increase (or decrease) in the density of a ubiquitous, natural, invisible, and mysterious mass-energy that permeates the supposedly empty space (the thing we call space-time) that the intrinsic accelerating motion of all matter in the Universe forms. So when we add accelerating matter to the Universe, additional mass-energy is concentrated from space around the matter and, at the same time, the matter itself contributes its own mass-energy to space through emissions of gravitational energy to help increase (or decrease) the density of this natural mass-energy[15]. Should the density go up, this is interpreted as the gravitational field getting stronger. If the density goes

---

15   The presence of a mysterious mass-energy permeating the entire Universe does conjure up the 19th-century idea of an aether. To many people, any discussion of an aether is usually a sign of crackpots doing second-rate science work. However, first-rate scientists such as Einstein never ruled out the idea. The problem that people have with accepting the idea lies in two important observations. The first has to do with the Michelson-Morley experiment. In that experiment, it was observed that the speed of light does not change with the motion of an object no matter which direction the light moves, and, therefore, it is assumed that an aether of the type that Sir Isaac Newton envisaged (basically having the mechanical properties of ordinary "solid matter fluids") could not possibly exist at absolute rest with respect to all other matter, let alone influence and help radiation to propagate through it. The second observation relates to the ability of high-frequency radiation to propagate through space. If the mass-energy of space were an ordinary mechanical fluid, then it would require this mass-energy of space to have the properties more akin to a piece of rigid steel. Yet somehow all of the planets, stars, and everything else in the universe must be able to "swim" in it. Again, such unheard of "magical" properties in a mechanical fluid along with the results of the Michelson-Morley experiment were enough to see the idea get thrown out, together with the term *aether*.

But have we prematurely thrown the baby out with the bathwater?

It should be noted that all of these observations do not necessarily disprove the existence of an aether-like substance permeating the Universe. For example, we know that radiation is energy, and it behaves like ordinary matter in many respects. The fact that this mass-energy of radiation permeates the Universe just helps to reinforce the aether-like nature of the substance. In fact, Einstein himself made it clear that his Special Theory of Relativity has not made the aether theory redundant. Rather, it is just that one may not speak of motion in relation to that aether. As he stated:

down, the gravitational field is said to be weaker. In other words, we are dealing with the density of the mass-energy that controls the strength of the gravitational field.

The best analogy for explaining this concept of density is to ask yourself which of the following two balls has more mass-energy and

"The Special Theory of Relativity forbids us to assume the [a]ether to consist of particles observable through time, but the hypothesis of [a]ether in itself is not in conflict with the Special Theory of Relativity. Only we must be on our guard against ascribing a state of motion to the [a]ether."

In general relativity, Einstein elaborated on this further. He said that space does have physical properties, and it can attribute physical properties to matter when it is curved (and so it can be considered to be not unlike the aether that a number of 19th century scientists described). As he once wrote:

"...space is endowed with physical qualities. [And] in this sense, therefore, there exists an [a]ether."

Even some physicists are looking into the possibility that dark energy may also fill all of space as an explanation for observations of an accelerating universe. If this is true, the idea certainly supports an aether-like substance.

What we are trying to say here is that just because one experiment has shown that nothing moving in space affects radiation, and thus, radiation does not need the type of aether that 19th century scientists described for its propagation, this does not mean that no energy exists in space. Of course, energy does exist in space. It is everywhere to be found. Nor can we assume that this)energy has no influence on the ordinary matter moving through space. For example, we know that radiation affects matter by moving it. So in a sense, radiation is a form of mass and energy, and it is a mass-energy substance that behaves in certain ways like an aether. Until we fully understand the physical properties of this mysterious mass-energy in space (which is probably just radiation), some people may think the space above the Earth's atmosphere does not contain energy or is simply too weak to affect ordinary matter and light. Not true. The reality is, there is energy in space, and it may very well have an enormous influence on many things that we have not realized, such as generating the so-called gravitational field and how protons stay together in the atomic nuclei.

At the present time, physicists are quite happy to replace the term "aether" with "fields" or "the fabric of spacetime". Nothing is wrong with the term "aether". It is just that scientists prefer to drop this word in favor of other, presumably more accurate, terms as a reflection of aether's connection to a specific set of 19th-century conceptual models that have now been disproven.

Perhaps it is time we change the name for this mass-energy in space to something else? As Hendrik Lorentz wrote:

"...whether there is an aether or not, electromagnetic fields certainly exist, and so also does the energy of the electrical oscillations [so that] if we do not like the name of 'aether', we must use another word as a peg to hang all these things upon."

How about we call this mass-energy of space "radiation"? Or, is there a more accurate name we can use?

hence more strength in the gravitational field: a tennis ball, or a cricket ball. With the two being roughly the same size, naturally you would go for the cricket ball. It is harder, and is filled with more mass-energy by way of solid materials compared with the gas-filled tennis ball with its rubbery and furry skin. So, therefore, the gravitational field generated by the cricket ball has to be stronger than the tennis ball's own gravitational field. Of course, this does not mean that the cricket ball will fall faster and hit the ground first if we let them go at a certain height. Although the Earth does attract or pull on the cricket ball more due to its increased gravitational field, the extra mass of the cricket ball also makes it more difficult to move, so it takes more force to move it. But when the cricket ball does move, it will fall at the same rate and hit the ground at the same time as the tennis ball, assuming no air resistance. At any rate, here comes the important bit. If you squeeze the tennis ball to such an extent that it becomes no larger than a marble, the density of the mass-energy in the tennis ball can be made to be high enough to generate a gravitational field that can equal or exceed that of the cricket ball. In other words, a compression has the effect of increasing the density of the mass-energy of space (and not just of the matter composing the tennis ball) as if the atoms in the compressed tennis ball are somehow accelerating at a greater rate (and, indeed, observations have shown an increase in the temperature at a higher density[16]), which, in turn, increases the strength of the gravitational field. Thus, without adding any extra mass to the tennis ball, a simple compression of the mass-energy will easily achieve the same effect of increasing the strength of the gravitational field for the entire tennis ball.

Why does this occur? It is because the movement of the subatomic particles and the atoms themselves, as well as their emission of extra mass-energy to the surroundings, becomes more agitated. It is also due to the close proximity of the accelerating particles and atoms to each other where they can pick up the higher-frequency mass-energy that this accelerating matter emits. This causes the acceleration of these particles to increase (e.g., vibrate more intensely), and this, in turn,

---

16   It is interesting to see a relationship between temperature—which is a measure of the quality (or
        frequency) of the radiation—and strength of the gravitational field. Is this a coincidence? We will have
        more to say about this when we discuss Einstein's Unified Field Theory in Chapter 3.

emits more higher-frequency mass-energy, resulting in a higher density of the mass-energy.

This is why in astronomy, the collapse of an ordinary star on itself to form a smaller object (e.g., a typical example would be a highly-dense neutron star) helps to create a more powerful gravitational field. It is all due to the fact that there has been an increase in the density of its own mass through compression, which, in turn, increases the density of the mass-energy permeating and surrounding this object. The atoms and subatomic particles making up the star get closer together, and so pick up the higher frequency mass-energy from each other, and then the particles simply accelerate faster. This higher acceleration helps to create a stronger gravitational field. Likewise, black holes are thought to be even more compressed and possibly faster-spinning (i.e., higher accelerating) stars, which is what gives these objects their legendary extreme gravitational fields that are capable of holding together millions of stars to form galaxies, as well as achieve the intensely high temperatures on the surface of the star from such high compression of matter.

And you thought a supergiant star capable of reigning in, say, the planet Mars and all of the other planets in the inner orbits of our solar system is massive compared with our mediocre sun. Forget size. You should try to compress it to the size of a neutron star less than 1,000 km in diameter. Then, you will definitely feel the extra mass through its powerful gravitational field. Not only that, but also if you should ever make the foolish mistake of choosing to land on the surface of a neutron star, you will definitely know what it is like to live in a two-dimensional Universe.

In summary, if you add extra accelerating atoms to an object, you can generate a stronger gravitational field. Or, if you remove the atoms from the object, you will decrease the strength of the gravitational field. Otherwise, you can compress the object to push the atoms closer together to create more acceleration and this will increase the strength of the gravitational field. Or, you can go the other way with the density to reduce the strength of the gravitational field. No matter what method you employ to control this accelerating motion of the atoms and subatomic particles, and no matter the direction of that acceleration, how much extra accelerating particles you add to something, or how much compression you apply, you can either increase or decrease the density of the mass-energy surrounding and permeating the object and this affects the strength of the gravitational field. It is this density that controls the strength of the gravitational field.

That is all the theory is trying to say to us.

We shall call this density concept the physicists' interpretation of the General Theory of Relativity as opposed to the mathematicians' interpretation we looked at earlier.[17]

But, of course, you might be wondering, what is this natural mass-energy we are talking about here? Einstein did not say anything other than how much of the mass-energy and its density controls the gravitational field. Certainly, this mass-energy is present in and around every object we see in the universe. It fills empty space and penetrates the very core of solid matter. It may even constitute the very nature of matter itself. Whatever it is, either the object is helping to generate and increase the density of the mass-energy by emitting extra mass-energy to the surroundings from the accelerating behavior of its own atoms and sub-atomic particles (and amplified by any acceleration that the object as a whole performs), or the object is somehow able to grab more of this mass-energy from its surroundings to amplify any inherent gravitational field that the object is generating. Or perhaps it is doing both? Of course, we still don't know what the gravitational field is. Whatever it is and how it gets created, the mass-energy of space must come with a gravitational field of its own for an object to be able to grab this energy and draw it to itself in a gravitational sense, and at the same time, the mysterious mass-energy must pull on the object in the same way (basically the familiar "rubber-band" effect we talked about earlier).

Whatever the mechanism behind all of this, the most that Einstein could say about this issue at the time that his paper describing the new theory of gravity was published is that the density of the mass-energy (or the curvature of space-time) somehow controls the strength of the gravitational field.

Beyond that, more work is needed to explain what this mass-energy is in reality.

---

17   Keep this "density of the mass-energy" idea in mind as this will come in handy when we make the interpretation of the Unified Field Theory in the next chapter.

One of sixteen original photographs taken by Sir Arthur Eddington on the island of Principe off the west coast of Africa during a six-minute period of the solar eclipse that commenced at 2:13 p.m. on May 29, 1919. This picture has faint line markings placed there by Eddington showing the positions of visible stars in the Hyades cluster. Another independent set of plates were taken at Sobral, Brazil, during the same solar eclipse. To have a control version, Eddington and his colleagues waited nearly six months to take another picture of the same stars. When the control plate was compared with the other plates, the scientists noticed the stars had moved from their expected positions near the Sun's domain from anywhere between 1.61 ±0.30 arcseconds (in Principe) to 1.98 ±0.12 arcseconds (in Sobral). As a result, Eddington and his colleagues had to conclude that Einstein's prediction for light bending in a gravitational field is correct and of a greater and more accurate magnitude as calculated by the gravitational field equations of Einstein's General Theory of Relativity (which predicted 1.75 arcseconds) compared to Newton's equations (which predicted 0.87 arcseconds).

The Gravitational Field Influences Light

It wasn't long before something intriguing started to emerge from Einstein's theory of gravity to make the great scientist look at light once again and to study it more closely.

As Einstein discovered, not only was density of the mass-energy important for controlling the strength of the gravitational field, but so too the speed and direction (or path) taken by the accelerating motion of all matter (including light) is affected by the density of the mass-energy as well. This is summed up perfectly by Professor John Archibald Wheeler when he said:

> "Matter tells space how to curve, and space tells matter how to move."

The best analogy we can give for this concept is to look at the behavior of an aircraft traveling through the air (a form of mass-energy). For example, we all know how an aircraft can travel in a straight line and at a roughly constant speed because the density of the air remains constant. But should the aircraft enter a pocket of denser air, the aircraft will suddenly change its speed and/or direction because the higher air density acts as a barrier to the aircraft's natural movement. The same is true of any matter traveling through the invisible mass-energy of the Universe according to Einstein's General Theory of Relativity.

And so too for light. Despite being a purely electromagnetic phenomenon, light is somehow able to bend its path in a gravitational field.

Come to think of it, the reason why scientists today accept Einstein's General Theory of Relativity is primarily because astronomer Sir Arthur S. Eddington (1882–1944) verified the qualitative correctness of the light-bending effect that the theory predicted using the tremendous mass of the sun and the position of light from a distant star near the sun's domain during a solar eclipse in 1919.

Yet despite realizing light bends in a gravitational field, Einstein discovered something odd. Why should light bend at all? As light is a purely electromagnetic phenomenon, a gravitational field should not affect it. And yet it does. The best that scientists can do to explain this phenomenon is to resort to Einstein's famous equation. In other words, although light has energy E, according to the famous equation $E=mc^2$, it must somehow produce a certain amount of mass m as well. And because it has mass (or acts as though it has mass even though scientists declare light to be massless[18]), it can be deviated by the mass-energy of space.

Yes, it sounds a bit confusing. But don't worry. Things will get a little clearer in the next chapter. For now, let us assume that light can bend in a gravitational field.

Still salivating for more insights? Okay.

As another consequence of the General Theory of Relativity, scientists have learned that the path that light takes actually bends more significantly in a high mass-energy density substance than in a low mass-energy density substance. Not only that, but also the speed of light in a region of high mass-energy density slows down. Thus, a dense mass-energy substance approaching that of a diamond will reduce the speed of light to 125,000 kilometers per second compared with 300,000 kilometers per second in the so called "vacuum of space" above the Earth's atmosphere. Increase the density of mass-energy and it is possible for light to travel very slowly (potentially at walking speeds if you like). Increase it a little more and you could stop light in its own tracks. Increase the density even further, and the light could be made to return on itself.

Now who says light cannot be recycled?

Likewise, go the opposite way by reducing the density of this mass-energy of space and you can get light to travel faster than 300,000 km/s. Should there be a zero mass-energy density (or true vacuum), it will correspond to infinite light speed. Evidence to support this claim can be found in the following article published in *New Scientist* on January 9, 2013:

"Light hits near infinite speed in silver-coated glass"
17:33 January 07, 2013, by Jeff Hecht

---

18   Scientists say that light has zero mass when at rest or when moving in a direction that you cannot detect. This seems to contradict the idea that if light hits something, it acts as if it has mass. The way in which to consolidate these two ideas is to imagine that you have a ship on the surface of the ocean. Place a weight machine underneath the ship and no mass will be detected. So you might say that the ship is massless. Similarly, light might appear massless while it moves and floats through the mass-energy of space, but if light hits solid matter, the light will display its mass.

A nano-sized bar of glass encased in silver allows visible light to pass through at near infinite speed. The technique may spur advances in optical computing.

Metamaterials are synthetic materials with properties not found in nature. Metal and glass have been combined in previous metamaterials to bend light backwards or to make invisibility cloaks. These materials achieve their bizarre effects by manipulating the refractive index, a measure of how much a substance alters light's course and speed.

In a vacuum the refractive index is 1, and the speed of light cannot break Einstein's universal limit of 300,000 kilometers per second. Normal materials have positive indexes, and they transmit at the speed of light in a vacuum divided by their refractive index. Ordinary glass, for instance, has an index of about 1.5, so light moves through it at about 200,000 kilometers per second.

No threat to Einstein

The new material contains a nano-scale structure that guides light waves through the metal-coated glass. It is the first with a refractive index below 0.1, which means that light passes through it at almost infinite speed, says Albert Polman at the FOM Institute AMOLF in Amsterdam, the Netherlands. But the speed of light has not, technically, been broken. The wave is moving quickly, but its "group velocity"—the speed at which information is traveling—is near zero.

As a feat of pure research, Polman's group did a great job in demonstrating the exotic features of low-index materials, says Wenshan Cai of the Georgia Institute of Technology, who was not involved in the work.[19]

Indeed, this mass-energy density can also explain why matter possesses discrete values. For example, a physics textbook will state that the rest mass for the electron is $9.10938291 \times 10^{-31}$kg. The reason for this rather precise value is due to the density of the natural mass-energy permeating the Universe, which reaches a specific value around the electron by its accelerating (i.e., spinning) motion. Change the mass-

---

19   http://www.newscientist.com/article/dn23050-light-hits-near-infinite-speed-in-silvercoated-glass.html.

energy density of space by raising or lowering it (or by accelerating or decelerating the spin rate of the electron), and the discrete value will change[20].

In the case of light, the mass-energy density also affects the speed by which this energy can travel.

This is already revealing some interesting new observations about light. More specifically:

1. Reducing the density of mass-energy in space will cause the light to bend less in a gravitational field, and not just increase its maximum speed;
2. Light lengthens its wavelength in a reduced mass-energy density environment;
3. A longer wavelength in light also reduces the electromagnetic energy density of the light; and
4. Lowering the energy density of light makes it less able to bend in a gravitational field.

We can't help noticing how light is behaving in a similar way to the mass-energy of space in terms of its electromagnetic energy density. Is there meant to be a connection between the two? Or is the electromagnetic energy of light merely responding to the changes in the mass-energy density of space?

The Problem of Gravity

So, it is clear. The problem of gravity has not yet been explained. Although the innovative new theory of gravity may have helped scientists to see for the first time the link between accelerating masses (or mass-energy density) and the gravitational field, in no way does it explain the gravitational field in its fundamental form.

You see, Einstein was still perplexed about two important aspects of his theory on gravity, which he needed to understand before he could provide a definitive answer.

First, Einstein did not have the faintest idea about how to relate this baffling mass-energy of the Universe to reality. It seems to pervade all

---

20   And with this change, even a tiny amount, can radically change the Universe we live in, and even who we are and what we look like. Many different universes can exist simply through the varying of the density of this mass-energy of space.

things, not to mention create all of the very things we see around us in what we call *matter*. At the same time, it somehow controls everything from the gravitational field to time, distance, direction, mass, charge and speed.

And second, even if we were to determine with absolute certainty the true nature and identity of this mysterious mass-energy of the Universe, could we determine how it actually creates the gravitational field in the first place? Or to put it bluntly, "What is the gravitational field?"

As perplexing as these problems were for Einstein, unfortunately, the General Theory of Relativity was not going to reveal the much-needed answers that Einstein was looking for. Why? It probably goes back to the quote from Einstein that mathematics does not relate well to reality. Mathematics is only as powerful as the picture that Einstein created in his mind, and his brilliant new picture of accelerating mass somehow controlling the gravitational field was just the start. More work had to be done in this area.

Light Moves Uncharged Matter

Another clue to have Einstein wondering about the nature of light concerns an experiment he performed in 1905 on the photoelectric phenomenon.

Upon reanalyzing the results of his experiment after 1917, Einstein realized that electromagnetic radiation was behaving in another peculiar way. Somehow, the more he looked at the results, the more he noticed how radiation was behaving—as if a stream of tiny particles or invisible matter was embedded in the radiation and, upon collision, caused the electrons to be ejected.

But it is not just these electrons that move in the presence of light. Light can also move uncharged matter, according to the work of William Crookes and his special invention.

In the late 18th century, William Crookes invented an instrument called a *radiometer*. The radiometer has four vanes attached to a needle inside of a glass vacuum flask. One side of each vane is coated with a black substance to help absorb light energy. But because the other sides of the vanes are silvery in color, light energy is absorbed *and* emitted, resulting in twice the force being exerted on the silvery side of each

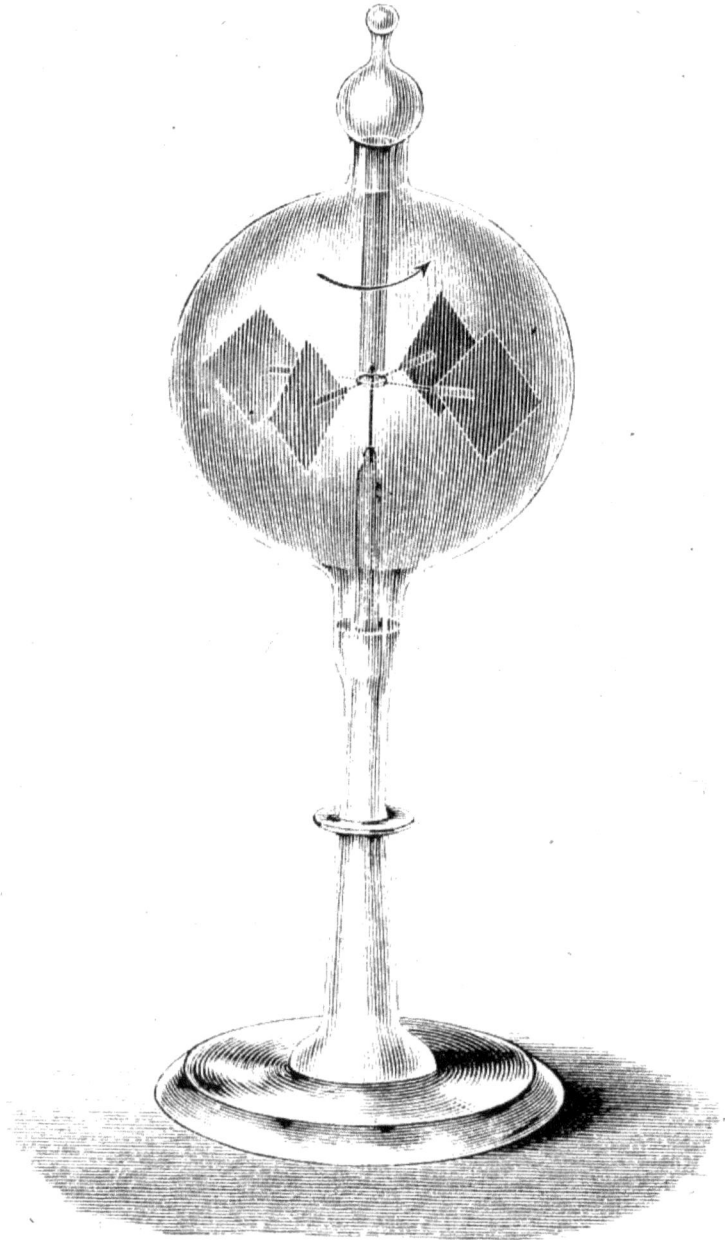

A radiometer demonstrates how uncharged matter moves in the presence of light

vane compared with the black side. In consequence, the vanes spin around on the tip of a fine needle when radiation is able to hit the surfaces of the vanes.[21]

This observation must have disturbed Einstein immensely because deep down he wanted to know how light moves uncharged matter. And, just like the gravitational field in the General Theory of Relativity, Einstein probably wanted to understand what is light in precise terms. Here we have two problems.

The problem Einstein had with light

It is clear Einstein had a problem with light.

At first, Einstein proposed in his photoelectric experiment that electromagnetic radiation really consists of bundles of concentrated electromagnetic energy called *light quanta* (or *photons*) and that this energy (depending on its density) is somehow responsible for the observed effect of charged or *uncharged* matter moving in the presence of radiation (or light).

Yet despite this explanation, Einstein was still not happy.

How could electromagnetic energy and its density affect radiation's ability to exert a variable force on uncharged matter? Surely there must be something immensely odd about this radiation.

As Professor Philipp Frank phrased it in his biographical analysis of the great man:

> "Einstein began to be much troubled over the paradoxes arising from the dual nature of light..."

Once again, what was a problem for Einstein was not one for the other scientists.

In the supposedly clear scientific language of the mathematicians and physicists of today, light is seen as pure electromagnetic energy composed of oscillating electric and magnetic fields at right angles to

---

21   The silvery side of the vanes will be pushed more significantly than on the black side so long as the temperature of the black surface does not exceed that of the silvery surface. To achieve this, spray the surface of the radiometer's glass flask with a refrigerant to extract the heat from inside. However, should the black surface be allowed to get hotter than the silvery side, it will radiate more of the lower-frequency radiation by way of heat into the environment. When combined with any remaining molecules in the near vacuum-like environment inside the glass flask, the flow of these molecules over the surfaces of the vanes with different temperatures can amplify the force exerted on the black surface. As a result of this heating-up effect, the vanes' rotation can be reversed.

each other and to the direction of motion. As far as the particle-like nature of light is concerned, scientists believe that we should think of it as a stream of compressed electromagnetic energy called *photons* traveling through space. Why it moves uncharged matter is best put down to the famous equation relating energy and mass. In other words, the fields somehow create a little bit of mass, and that's sufficient to explain it. But don't go further. There is no need to bust your brains over this seemingly trivial problem.

As blindingly clear as this description of electromagnetic radiation, or light, in its most general sense, might seem to be to scientists, it is still not terribly enlightening to the average person on the street. Seriously, how many readers could attest that they can visualize oscillating electric and magnetic fields in space being bunched up to form things called photons and somehow causing motion in charged and uncharged objects? Clearly, a simpler and less mathematically inclined description is required.

Einstein probably thought the same way, too, at some point. The complicated description of light was not helping the matter for Einstein because he felt that light was hiding another issue. Yes, he knew the density aspect remained somehow important to this whole debate. But for some reason or another, Einstein seems not to have been fully satisfied. He must have wondered to himself, "Exactly how could light (or the photon), a purely electromagnetic phenomenon, affect an uncharged object through energy density when it is known, according to the classical laws of electromagnetism, that it should affect only charged objects?"

This suggests that Maxwell's law of electromagnetism cannot be complete. Well, let us put it this way: Maxwell certainly made no mention in his electromagnetic theory of how radiation can affect uncharged matter, let alone bend in a gravitational field. Einstein understood this very well. After a while, he took the next logical step of thinking that Maxwell had not completed his work to take into account the effect of light on uncharged matter[22]. But to complete the work, Einstein had to make the decision, as many scientists do today, to think that the gravitational field is a real and distinct force of nature. Or to

---

22   This is not because Maxwell chose not to complete his work. He was not aware of the situation between uncharged matter and radiation. His focus was to show that radiation was an electromagnetic wave. He died in 1979 at the young age of 48 and had very little time to notice and investigate the effect of radiation on William Crookes' radiometer (which was invented in 1873).

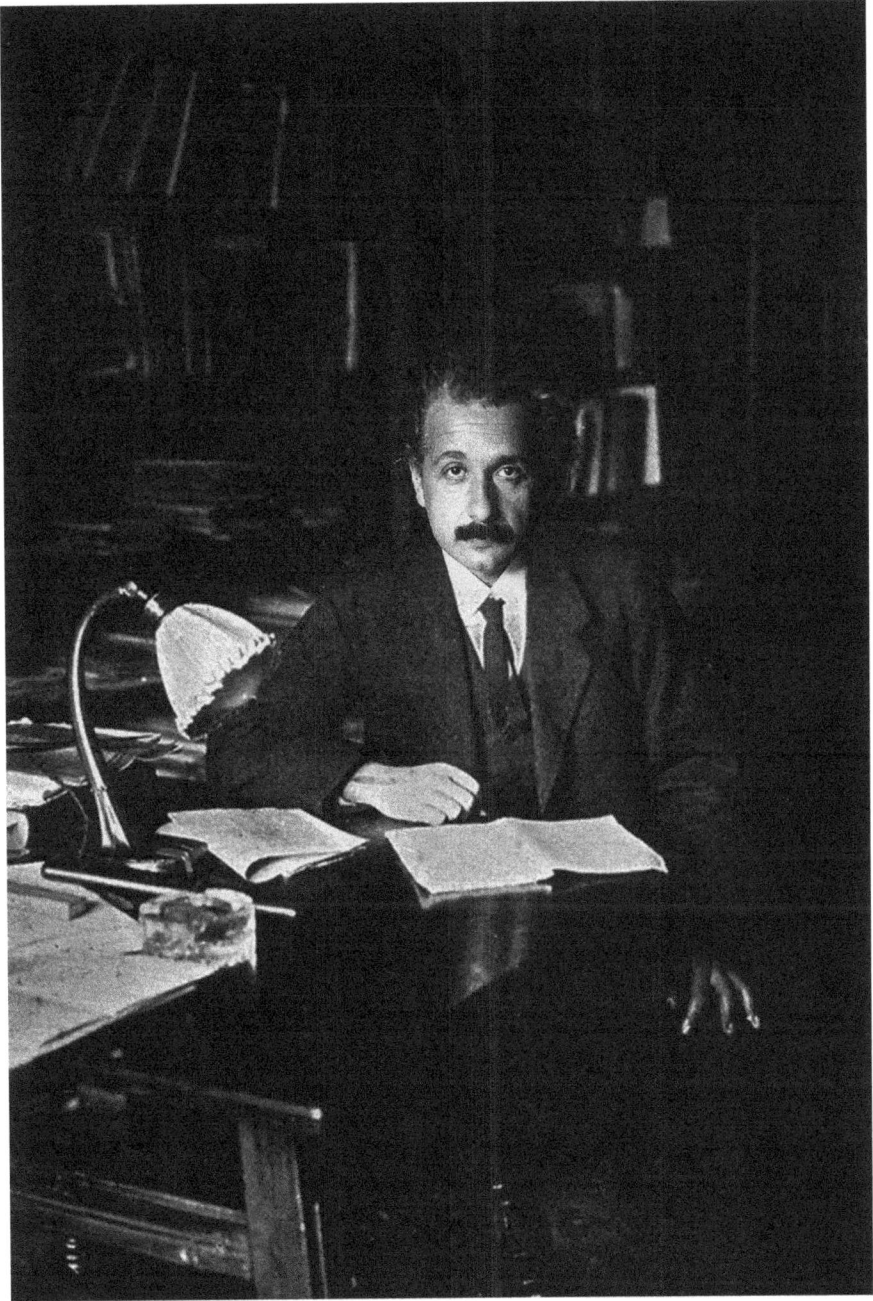

Albert Einstein in his office at the University of Berlin in 1920.

put it another way, no connection allegedly exists between gravitational and electromagnetic fields. Both must be independent based on current scientific knowledge.

Similarly, Einstein's General Theory of Relativity was not complete either, because it could only explain uncharged objects in an accelerated frame of reference.

So, what did Einstein do next?

Assuming that Maxwell's law of electromagnetism is not complete, somehow Einstein made the decision to combine electromagnetism with the General Theory of Relativity so that he could show how light can affect uncharged matter.

The Unified Field Theory Comes of Age

Einstein worked diligently after 1919 to develop his new electromagnetic/gravitational or quantum/classical theory—or the mother of all scientific theories, as some scientists would like to call it. He would coin the term the "Unified Field Theory" to describe the formidable work he did in this area.

The first indications of how close he was to succeeding in his unification work came in 1922 when Einstein tentatively published a 6-page paper containing the results of his ambitious work in the *Proceedings of the Academy of Sciences*, although he later withdrew the paper on the grounds that it was incomplete[23].

Einstein took a new approach to the problem, this time using what he termed "distant parallelism" in Riemannian geometry. However, due to the number of calculations involved in this new approach to help him to complete his unified field work, and in order for him to concentrate on accomplishing the work at some point in his life, Einstein had to complete and publish another paper in 1924, where he explained his last important discovery relating to the association of waves and matter.

Several years passed until, by the summer of 1928, he had published an improved version of his theory titled "Neue Möglichkeit für Eine Einheitliche Feldtheorie von Gravitation und Elektrizität"[24] (translated

---

23   Berlitz & Moore 1979, p.140.
24   Published in Volume 18 (1928) in the journal *Verlag der Akademie der Wissenschaften in Kommission bei Walter de Gruyter* (translated as "Publisher of the Academy of Science in Commission at Walter de Gruyter") located in Berlin.

as "New Possibility for a Unified Field Theory of Gravitation and Electricity"). And on January 10, 1929, he published a completed 8-page version[25] in the same journal.

News of the completed work soon reached the press with the *London Times* of February 4, 1929, quoting a translation by L.L. Whyte from a scientific source:

"This paper represents a new development which was immediate news."

As Professor Banesh Hoffmann of the University of New York and Helen Dukas (1896–1982), Einstein's secretary from 1928 until his death in 1955, noted in their book *Albert Einstein: Creator and Rebel*:

"In 1928, [Einstein] embarked on a new approach to a unified field theory...involving what he called "distant parallelism"...By early 1929 he had solved the main problems involved in writing down field equations for his unified field theory. On the day of official publication of the third of a formidably technical series of nine articles on the theory... excited headlines appeared in foreign newspapers throughout the world....In this frenzied, unscientific atmosphere, Einstein's new theory was hailed in the press as an outstanding scientific advance. Yet Einstein had stated in his article that this was still tentative..."[26]

Later in the same year, Professor Tullio Levi-Civita of the University of Rome produced a paper "suggesting ways in which, in his opinion, the presentation of the theory could be simplified and improved"[27]. Levi-Civita explained his method of simplifying Einstein's unifying work by saying:

"It appears to me, however, that the root problem raised by Einstein can be solved in a simpler and more general way by making use of perfectly familiar methods of the absolute

---

25   Published in Volume 1 (1929) of the same Prussian scientific journal as his 1925 article.
26   Hoffman & Dukas 1972, pp.225-226.
27   Levi-Civita 1949, p.0 (cover page).

differential calculus on the one hand, while, on the other hand, retaining unaltered all results previously obtained."[28]

Einstein agreed.

It wasn't long before Einstein simplified his unified field equations after applying Hamiltonian principles to the problem. Einstein quickly republished his latest version of the Unified Field Theory again in 1929 under the title "Einheitlichen Feldtheorie Und Hamiltonsches Prinzip" (translated as "Unified Field Theory and Hamiltonian Principles") in the same Prussian journal.

Yet the problem of relating the mathematics of his unified field equations to the real world was only beginning to dawn on Einstein. As science historian and author Abraham Pais (1918–2000) said:

> "[Einstein may have] proposed a set of field equations, but added that "further investigations will have to show whether [these] will give an interpretation of the physical qualities of space."[29]

By around mid-1929, Einstein decided it was time to start deriving solutions from his Unified Field Theory and see whether they could be related to reality, and hopefully, therefore, be shown to be valid through observations or an experiment. Only one thing was bugging him: how long it would take to write out all of the calculations. It had already taken him many years to complete his Unified Field Theory. His only hope now was to get assistance from other people who knew something about the mathematics he was using.

Professor Walter Mayer, a competent mathematician by trade, was asked to take care of writing out the calculations on Einstein's behalf. He agreed. Einstein allegedly said about Mayer:

> "It is he who produced all my calculations; his skill is fantastic, you know."

Einstein's relationship with Mayer proved useful. Evidence of this can be seen when Einstein and Mayer succeeded in deriving two "static solutions".

---

28   Levi-Civita 1949, p.3. Levi-Civita's paper was translated into English and published in pamphlet form by Blackie & Son Limited in London.
29   Pais 1982, p.346.

The paper describing a couple of relatively simple "static solutions" appeared in 1930. The 11-page paper was titled "Zwei Strenge Statische Losungen der Feldgleichungen der Einheitlichen Feldtheorie" (translated as "Two Strict Static Solutions to the Field Equations of the Unified Field Theory"). But the real hard work was yet to come. Einstein still had to attempt the next ambitious step of looking at non-static field solutions. It is believed that these "non-static field solutions" would be crucial to understanding how light can move uncharged matter and perhaps explain the mystery of gravity as well.

In Spring 1931, Einstein and Mayer produced a second paper relevant to the Unified Field Theory. It was titled "Sytematische Untersuchung über Kompatible Feldgliechungen, welche in Einem Riemannschen Raume mit Fernparallelismus Gesetzt Werden Konen" (translated as "Systematic Study of Compatible Fields, Which Are Set in Riemannian Space with Distant Parallelism").

After spending another couple of years diligently working on the Unified Field Theory, Einstein decided to travel abroad to England, Asia, and the United States. It was in England that Einstein gave his first lecture in English, to an eager audience at the George A. Gibson Foundation at the University of Glasgow. There, he explained the origins of the General Theory of Relativity, giving the academics who were present a quick crash course on Riemannian geometry for anyone who felt a little rusty in this area (don't we all!). Then, there were those who didn't care so much about the mathematics but just wanted to be there to see the great man.

Later, with his wife Elsa and sister Maja by his side, he visited the United States where he received a warm welcome from the media and the public. Soon, American scientists were clamoring for an opportunity to listen to this great man and pure intellectual genius talk about his work.

Suddenly, the political tensions in his homeland of Germany were escalating, much to Einstein's great personal sadness. Many German people were looking for a solution to their country's worsening economic climate. And one political party was prepared to make one group of people living in the country the scapegoat for all of the problems: the Jews.

As a Jew and academic, Einstein knew his life would be endangered if he returned home. With his parents and uncle now dead and his sister and wife already out of Germany, the only thing he had to do was find a safe haven. Fortunately, it turned out that the United States was willing to offer Einstein a permanent place at the then-newly established Institute for Advanced Studies at Princeton University, New Jersey, where he was to stay quietly for the remainder of his life. He assumed the post in 1933, and by 1934 he had received his American citizenship.

The house where Einstein lived for the rest of his life.

At the height of the Second World War, the Unified Field Theory did resurface in another scientific publication sometime in 1940. After the war, Einstein published again in December 1949, with the latter publication revealing "Professor Einstein's most recent extension of his Unified Field Theory"[30]. According to publishers Blackie & Son Limited who reprinted Levi-Civita's paper in December 1949:

> "Since then [1929] and until December 1949 there appears to have been no major development in this field of study."

---

30   Levi-Civita 1949, p.0.

In 1943, a couple of people from the U.S. Navy's research division may have stumbled onto something in relation to Einstein's final great theory, and an experiment may have been conducted under great secrecy to show how an electromagnetic field can be used to render a ship invisible. Whether or not this is true, two U.S. Navy officers had their photo taken with Einstein in his Princeton home as a memento and asked Einstein to assist in short-term contract work on high explosives.

But then, after 1949, Einstein decided not to talk much about the Unified Field Theory. Any further updates he may have made to the Unified Field Theory were not publicized. Rumors were floating around that Einstein had destroyed some papers—presumably those having to do with his Unified Field Theory—only a few months before his death:

> "One such story has it that some months before his death Einstein had burned papers relating to some of his more advanced theories on the grounds that the world wasn't ready for such things and would be better off without them."[31]

This is not surprising considering that Einstein lived through two world wars, saw the destructive application of his famous equation linking mass and energy through the development of atomic bombs dropped onto two Japanese cities, and witnessed the beginning of the Cold War between the nuclear powers of the United States and Russia. The work for the U.S. Navy was not helping Einstein to see a positive future for the human race and one filled with hope. All of this was definitely not to his liking. Einstein's view on the abhorrent nature of war is evident when he said:

> "He who joyfully marches to music rank and file, has already earned my contempt. He has been given a large brain by mistake, since for him the spinal cord would surely suffice. This disgrace to civilization should be done away with at once. Heroism at command, how violently I hate all this, how despicable and ignoble war is; I would rather be torn to shreds than be a part of so base an action. It is my conviction

31   Berlitz & Moore 1979, p.143.

that killing under the cloak of war is nothing but an act of murder."[32]

And in another statement, he said:

"Peace cannot be kept by force; it can only be achieved by understanding."[33]

Thus, it is altogether possible that Einstein had discovered something significant and had decided to keep it a secret because of the tendencies toward war of his fellow human beings. And to make sure of that, he could have made others believe that he failed when he said to his friend Michael Besso in 1954:

"All these fifty years of conscious brooding have brought me no nearer to the answer to the question, 'What are light quanta?' Nowadays every Tom, Dick and Harry thinks he knows it, but he is mistaken. ... I consider it quite possible that physics cannot be based on the field concept, i.e., on continuous structures. In that case, nothing remains of my entire castle in the air, gravitation theory included, [and of] the rest of modern physics."[34]

Or, it may simply be an admission of Einstein's failure to reach an ultimate solution and, consequently, his unifying attempts were unsuccessful. However, as history tells us, Einstein continued to stick by his final theory, even on his death bed. Such unwavering certainty in his work does not suggest that he was getting nowhere. Far from it. He was confident about something. We get a glimpse of this confidence when we see the following rather interesting statement he made in his letter to Besso:

"...the idea that there exist two structures of space independent of each other, the metric-gravitational and the electromagnetic, was intolerable to the theoretical spirit. We are prompted to the belief that both sorts of field must correspond to a unified structure of space."[35]

---

32   From Einstein's book, *Mein Weltbild* (*My World-view*), published in 1931.
33   From a speech to the New History Society, December 14, 1930.
34   In a letter written by Einstein to his old friend Michael Besso, 1955.
35   In a letter written by Einstein to his old friend Michael Besso, 1955.

In other words, he felt confident that the gravitational and electromagnetic fields had to be combined into a single unified structure. He knew beyond any reasonable doubt that this structure had to be an intrinsic part of space-time. Nothing was leading him to question this aspect. Yet deep down, Einstein was still grappling with something more fundamental than this. More specifically, it was probably a question of how to separate the gravitational field from the electromagnetic field in his own mind so that he could properly explain what the gravitational field is in its true sense. Sure, his mathematics could make this separation clear on paper, but the equal sign designed to show the mathematical link between the fields did not help him to get to the heart of the matter—namely the true nature of the gravitational field.

Could this be the reason that Einstein spent so much time on his final theory?

Then, not long after he gave the above quotes, Einstein decided to go back to his theory. Why? Part of the reason might be because Einstein knew that his time was coming to an end after learning from his doctors that surgery was recommended to again fix his ailing heart, except Einstein refused. So, although he knew his body was failing., he may have chosen to apply himself one more time to the problem. However, the more likely reason is because he wanted to say something to the world. This is evident by the fact that Einstein did not want to state his work on the Unified Field Theory was a failure. Something made Einstein confident that he was on the right track. Does this mean Einstein saw a way out of this unified structure problem by finally resolving the true nature of the gravitational field just prior to his death but was afraid to tell others what he saw? Or, was he trying to let people know indirectly that the unified structure is not only correct, but also should be pursued to a logical conclusion for us to properly get to the end and make our own discovery? Without a thorough investigation of his final theory, we can only speculate. But following the death of his wife, his sister and all of his remaining close friends from his childhood, Einstein lived and worked a mostly secluded life in his rented house on 112 Mercer Street, Princeton, about three-quarters of a mile from the Institute[36]. In spite of the sense of isolation that all of

---

36   Bernstein 1991, p.11.

this must have generated for him, his love of mathematics, philosophy, music, and sailing, as well as the beauty of watching nature at work still kept his mind young and active, even to his final day.

We see his active mind on April 17, 1955—the day before his death—when he phoned his secretary, Miss Helen Dukas[37] from his hospital bed, asking her to bring "my most recent calculations" concerning the Unified Field Theory. Perhaps he did this in the hope of finishing his work or maybe to let us know something about the importance of his work to science.[38]

Einstein, the greatest physicist of the 20th century, would leave behind his two sons—Eduard (1910–1965) and Hans Einstein (1904–1973)—from his former marriage to Mileva Maric (1875–1948).

But why spend so many years valiantly pursuing such an ambitious theory of unifying physics? Is there anything we can learn from Einstein's final theory?

---

37   Bernstein 1991, p.5.
38   Einstein had been working on a draft speech to mark the anniversary of Israel's Independence Day and to express his concerns about the long-term conflict between Israel and Egypt. He never completed his speech. He stopped at the end of this paragraph when he realized he needed to rest:

> "You may think this is a small and insignificant problem and that there are more serious things to worry about. But this is not true. In matters of truth and justice there can be no distinction between big problems and small...Whoever is careless with the truth in small matters cannot be trusted in important affairs...."

Despite resting in his home, Einstein knew that his health was deteriorating. He was taken to Princeton hospital on April 17, 1955, in a wheelchair. Doctors who examined him acknowledged that he had suffered an internal bleeding due to aortic aneurysm rupture. It was a condition that had been treated before. However, on this occasion, Einstein refused to have surgery, stating to his doctors:

> "I want to go when I want. It is tasteless to prolong life artificially. I have done my share, it is time to go. I will do it elegantly."

Knowing that his time was almost up, he asked his secretary, Helen Dukas, to bring him the last calculations he had done on the unified field equations on April 17, 1955. He received the papers and was left alone in his hospital bed to spend a few more hours working on his theory. Even though he knew any ideas he had come up with or solutions in his mind of how to get to the final answer for the gravitational field would not be revealed to help scientists experimentally verify the validity of his final theory, his continuance in working on his theory right to the end was probably more a statement to the world from his deathbed that what he had achieved was correct. All it requires are others to find out what this secret is.

A nurse later arrived to check on him, and it was about this time that he stated to his nurse, "...I think I will rest for a while." He placed his papers and pen on his bedside table and went to sleep.

He never woke up again. Early the next morning on April 18, 1955, with his son Hans Albert at his bedside, Einstein died at 1:15 a.m.

Apart from the pathologist at the hospital who made the decision to remove and preserve Einstein's brain for study, reportedly without the permission of his family, the rest of the body was cremated on the same day and his ashes were scattered at an undisclosed location not far from the Delaware River, following Einstein's wishes.

As for Einstein's unified field papers, souvenir hunters apparently stole those left behind on the bedside table. Of the remaining papers found in his Princeton home, no indication existed that Einstein had doubts about any aspect of his unified field work. Einstein had always maintained that what he had done was right.

CHAPTER 3

The Interpretation of the Unified Field Theory

Logic will get you from A to B. Imagination will take you everywhere.

—Albert Einstein[1]

EINSTEIN's Unified Field Theory, also known as the Einstein-Maxwell theory, was first published in a Prussian scientific journal in 1922, with the final completed version in 1929. But as Graham Keith Russell queried in his 1972 thesis titled *The Interpretation of Einstein's Unified Field Equations*:

> "How do the field equations describe the geometry of space-time, and how do they describe the electromagnetic and gravitational fields in that space-time?"[2]

---

1   Origin of this quote is unknown, but is published in *The Ultimate Quotable Einstein* (2010) by Alice Calaprice and Freeman Dyson on p. 481 with the proviso that it is attributed to Einstein considering he has made numerous statements supporting the importance of imagination to his own work.

2   Russell 1972, p.94.

To the physicist whose job it is to find realistic solutions and with it the possibility of conducting an experiment to prove the theory, the question is whether there is anything in the real world to support the relationship between the electromagnetic field and the gravitational field.

Zur einheitlichen Feldtheorie.

Von A. Einstein.

In zwei jüngst erschienenen Abhandlungen[1] habe ich zu zeigen versucht, daß man zu einer einheitlichen Theorie der Gravitation und Elektrizität dadurch gelangen könne, daß man dem vierdimensionalen Kontinuum außer einer Riemann-Metrik noch den »Fernparallelismus« als Eigenschaft zuschreibt. In der Tat gelang es auch, dem Gravitationsfelde und dem elektromagnetischen Felde eine einheitliche Deutung zu geben. Dagegen führte die Ableitung der Feldgleichung aus dem Hamiltonschen Prinzip auf keinen einfachen und völlig eindeutigen Weg. Diese Schwierigkeiten verdichteten sich bei genauerer Überlegung. Es gelang mir aber seitdem, einen befriedigenden Weg zur Ableitung der Feldgleichungen zu finden, den ich im folgenden mitteile.

§ 1. Formale Vorbereitungen.

Ich benutze die Bezeichnungen, welche neulich Hr. Weitzenböck in seiner Arbeit über den Gegenstand vorgeschlagen hat[2]. Die ν-Komponente des s-ten Beins des n-Beins wird also mit h^s_ν bezeichnet, mit $^s h_\nu$ die zugehörigen normierten Unterdeterminanten. Die lokalen n-Beine sind alle »parallel« gestellt. Parallele und gleiche Vektoren sind solche, welche — auf ihr lokales n-Bein bezogen — gleiche Koordinaten haben. Die Parallelverschiebung eines Vektors wird durch die Formel

$$\delta A^\nu = -\Delta^\nu_{\alpha\beta} A^\alpha \delta x^\beta = -s h^\nu\, {}^s h_{\alpha,\beta}\, A^\alpha \delta x^\beta$$

gegeben, wobei in $^s h_{\alpha,\beta}$ das Komma andeuten soll, das nach x^β im gewöhnlichen Sinne differenziert werden soll. Der aus den (in α und β nicht symmetrischen) $\Delta^\nu_{\alpha\beta}$ gebildete »Riemannsche Krümmungstensor« verschwindet identisch.

Als »kovariante Differentiation« verwenden wir nur jene, welche mittels der Δ gebildet ist. Sie sei nach der Gepflogenheit der italienischen Mathematiker durch ein Semikolon bezeichnet, also

$$A_{\mu;\,\tau} \equiv A_{\mu,\,\tau} - A_\alpha \Delta^\alpha_{\mu\tau}$$
$$A^\mu{}_{;\,\tau} \equiv A^\mu{}_{,\,\tau} + A^\alpha \Delta^\mu_{\alpha\tau}$$

Da die $^s h_\nu$ sowie die $g_{\mu\nu}$ ($\equiv {}^s h_\mu {}^s h_\nu$) und die $g^{\mu\nu}$ verschwindende kovariante Ableitungen haben, können diese Größen als Faktoren mit dem kovarianten Differentiationszeichen beliebig vertauscht werden.

[1] Diese Berichte VIII. 28 und XVII. 28.
[2] Diese Berichte XXVI. 28.

First page of the original 6-page paper for the Unified Field Theory published in 1929

Of course, we could always let the mathematicians run rampant on the Unified Field Theory by letting them make a detailed mathematical analysis of the field equations using absolute differential calculus, see what sort of solutions[3] emerge from their efforts, provide hopefully clear descriptions of what these solutions are, and then later try to relate these solutions with reality. Then again, given how well mathematicians have been able to explain the General Theory of Relativity using the "curvature of space-time" concept in the previous chapter, maybe this isn't such a good idea. And anyway, letting our beloved mathematicians do the work of understanding the Unified Field Theory would only seriously complicate this book for the average reader.

Seriously, what are the chances of a mathematician finding a solution and a suitable real-life interpretation after undertaking a mathematical approach to this problem? As history has shown, other scientists have tried this direct and rather complicated mathematical approach and all have given up. This is confirmed in the preface of Mrs. M. A. Tonnelat's 1966 book, *Einstein's Unified Field Theory*, where André Lichnerowicz writes:

> "As a legacy, Einstein has left an enigmatic theory that scientists view with suspicion and hope. A large amount of work is necessary to either prove or disprove the theory...."[4]

Tonnelat herself agrees with this view, saying:

> "...The unified field theory continues the simplicity of its principles with a profusion of calculations and a wealth of formalism."[5]

Because of this "wealth of formalism" and the need for the near-equivalent to brain surgery in performing the "profusion of calculations" required to understand and support the Unified Field Theory, it seems that by the early 1960s most if not all scientists had pretty much chosen to deny the possibility that Einstein may have been successful.

---

3   The same thing is being done with the General Theory of Relativity with mathematicians coming up with extraordinary mathematical solutions, such as the possibility that wormholes in space known as the Einstein-Rosen bridge might exist in space. However, many of these solutions have yet to be shown as having a true basis in reality.

4   Tonnelat 1966, p.v (Quote from Lichnerowicz).

5   Tonnelat 1966, p.vii.

As Dr. Cornelius Lanczos of the School of Theoretical Physics at the Dublin Institute for Advanced Studies has summed up the situation:

> "In the meantime, "modern physics" continues to grow and advance without taking account of Einstein's unifying attempts and, in fact, denying even the possibility of such an attempt being successful."[6]

Clearly there has to be another way.

Was Einstein Wrong?

Of course, there is always the possibility that Einstein was getting nowhere with his final great theory. The highly complicated nature of the mathematics may have led Einstein up the proverbial garden path so to speak. As American theoretical physicist Michio Kaku said, in reflecting much of the view of the scientific community on Einstein's final legacy:

> "Einstein always began with the simplest possible ideas, and then put them into their proper context. But Einstein failed in his attempt to create a unified field theory because he abandoned this simple conceptual approach and instead resorted to the safety of obscure mathematics."[7]

The only thing is, Einstein showed no signs of giving up on his theory, even right up to the end of his life. As we have seen from the previous chapter, Einstein asked his secretary to bring in his latest calculations on the Unified Field Theory even though he knew his time was running out. Spending several more hours working on his theory while sitting on his deathbed is not exactly clear evidence of his willingness to admit defeat in this area. Something told Einstein to pursue his ambitious dream.

So despite the pessimistic views of Einstein's contemporaries, the Unified Field Theory, as committed to paper by Einstein, cannot be dismissed as the wishful ramblings of an old man. Neither can we assume that the unified field equations have no bearing on the real

---

6   Lanczos 1965, p.118.
7   http://www.philosophymagazine.com/extra/quotes_bydate.htm

world just because some scientists find it too difficult to uncover solutions from a purely mathematical approach. How do we know for sure that Einstein's original "picture" he had for light that made him pursue his final theory is not correct?

In fact, we should be asking the scientists, What was the idea that Einstein uncovered that made him convinced he had to unify the gravitational and electromagnetic fields into a single unified field? And could this idea have a relationship to some known natural phenomenon?

It may surprise readers to know that no one in the scientific community knows what idea(s) may have inspired Einstein to write his final theory. Most scientific efforts to understand Einstein's work have focused exclusively on a mathematical approach to finding solutions. And as Tonnelat has stated, that task in itself is horrendous due to the sheer amount of work to solve the equations.

But even if a solution could be found, there is also the problem of how to relate the solution to reality. Yet another hurdle for the weary scientist to overcome.

The only thing scientists can say with certainty about Einstein's work is that his mathematics and derivation of the final unified field equations is correct. All mathematicians currently support Einstein's unified field equations, claiming them to be a masterful intellectual achievement as Einstein had them published in his 1929 paper. Einstein clearly knew what he was doing. Furthermore, he did the work because of some serious attempt to solve a problem associated with the "paradoxes arising from light". His determination to solve this issue relating to light suggests that Einstein was confident that the electromagnetic and gravitational fields had to be mathematically cemented into a unified field. Clearly, Einstein saw something; otherwise, he would have declared the work to be flawed or a failure (and certainly well before he died in 1955). But he didn't, not even on his deathbed.

The only thing we can say is that Einstein kept mentioning "light" (or electromagnetic radiation) as somehow being important to his theory. Does this mean light is the thing we need to focus on when understanding whether the theory has anything to relate to the real world?

Is It Time to Face Reality?

As Einstein was fixated on the nature of light, let us begin by exploring more closely this interesting phenomenon. The aim here is to develop a clear picture in our minds of what it is Einstein was trying to solve and what he saw in his own mind that eventually led him to develop his Unified Field Theory.

Don't worry about the mathematics. There is still hope for us mere mortals in understanding how Einstein arrived at his theory and what he saw in it that made him so certain he is right. How? Simply by not following the exact same mathematical path Einstein and other scientists had taken. In fact, we can approach the problem in a different way.

Not sure how?

Perhaps it may come as a surprise that the aim of all good scientists, especially the physicists, is to simplify everything, including mathematics. How? By using the power of visualization and thinking to create a clear *picture* in our minds of what is at the heart of the problem. Ignore Kaku's claims that Einstein lost his way by not having a simple idea to support his mathematics. This may be untrue. Einstein may well have created the idea and formulated a simple picture in his mind before he even committed a single mathematical symbol to paper. It is up to us to find out what the picture might be and, with it, the idea he had all along. To do that, we must apply visual thinking skills in the same way.

Physicist Dr. Kent Cullers understood the importance of creating a picture in our minds through visual thinking skills before applying mathematics when he said:

> "As a blind physicist, surprisingly enough I spend a great deal of time doing the equivalent of visualizing because if you don't have a picture of the situation, you have no idea whether the answers that come from your equations make sense."

That is, once you have visualized the picture, the answer will often become self-evident well before the mathematics is applied to the problem.

Furthermore, the picture has to be one that clearly supports the theory, and a good one at that too. Mathematics is only as good as the picture you have created for it. So if the picture is not good in the first place, all the mathematics in the world will not help you to get solutions that makes sense and relate closely to reality. It is essential that we find the correct picture that made Einstein so incredibly confident in his work.

Finally, no matter how good a picture you create, the solutions derived from the equations must be testable. This is the final and most important part of the scientific process, probably more so than applying mathematics. The mathematics is only there to reinforce and give scientists greater confidence that the solutions obtained and the picture formulated in the mind to explain those solutions are probably right and are likely to relate to reality. But all the best and well-intentioned pictures and complicated mathematical solutions created by the best scientific minds can still fall apart if real-world experimentation does not support the picture or the solutions. This is the only way to determine whether the picture you have created and the subsequent mathematical solutions you derive and can explain through your picture are likely to exist in reality, and with it give greater credence to your picture.

However, before we get to this point, and especially given the mathematical complexity of Einstein's final theory, it is important to get the picture right first. It means we must visualize the problem and uncover the essential picture Einstein saw when he created his theory. Once we do that, there is an excellent chance we may also know the answer that Einstein had found that made him pursue his ambitious theory.

The Search for the Interpretation Begins

To begin our quest to find this elusive picture, let us be absolutely clear about one thing: the aim of Einstein in developing his Unified Field Theory was nothing more than to link the electromagnetic field with the gravitational field. No great Harry Potter magical quest on the part of Einstein. It is plain and simple physics. Why? Because Einstein felt confident that the two fields had to be linked and developed a somewhat complicated mathematical superstructure to support his thinking.

Tonnelat would agree with this view when she defined the unified theory in the following manner:

> "We shall say that a unified theory is one which will unite the electromagnetic and the gravitational fields into a single hyperfield [or unified field] whose equations are the conditions imposed on the geometrical structure of the universe."

To see what we mean by this mathematical linking of the electromagnetic field with the gravitational field, let us look at the structure of the unified field equations as published in advanced university textbooks on Einstein's relativity theory. Yes, we can hear some readers rushing out the door in a state of panic knowing they will be forced to peek at some dreaded mathematics. But remember, we will not be analyzing the equations too deeply. Indeed, this will be nothing more than a general look at how the unified field equations have been structured. If you can understand the straight line equation $y=mx+b$ from your high school days, then you will understand the structure of the field equations.

Are you ready? Okay, let's spill the beans on the equations.

The unified field equations

The unified field equations of the Unified Field Theory for the four-dimensional space-time world consisting of sixteen second-order partial differential equations are given in short vector notation form by:

$$\mathbf{G}_{u,v} = k.(\mathbf{T}_{u,v} + \tau_{u,v}) \text{ where } u,v = 0,1,2 \text{ and } 3$$

The terms $\mathbf{G}_{u,v}$, $\tau_{u,v}$, $\mathbf{T}_{u,v}$ and k are defined as follows:

$\mathbf{G}_{u,v}$ is described by mathematicians as a tensor (displayed in vector form by its **bold** styling) representing the ten field equations of gravitation for the four-dimensional space-time world, called the *gravitational field tensor* or *Einstein tensor*.

$\tau_{u,v}$ is another tensor (in vector form) representing the six field equations of electromagnetism for the four-dimensional spacetime world, called the *electromagnetic field tensor* or *Maxwell tensor*.

$T_{u,v}$ is the third (and final) tensor (in vector form) representing the mass or energy content of space, called the *external energy-momentum tensor*.

k is a constant that includes the gravitational constant equal to 6.673×10^{-11} N.m^2.kg^2 and the speed of light equal to 300,000 kilometers per second.

Are you already having a brain stroke? Relax. Don't look too hard into the equations. It really is very simple. We have even made it simple by displaying just the basic vector notation format through a couple of letters and a Greek symbol (i.e., **G**, **T** and τ) when simplifying the hideous tensor equations, exactly as Professor Tullio Levi-Civita did when he presented to the world the simplified version of Einstein's unified field equations. Just keep in mind the general structure of the equations in vector form as seen above because we will now compare them with the gravitational field equations of the General Theory of Relativity.

The gravitational field equations

Here are the gravitational field equations of the General Theory of Relativity for the four-dimensional space-time world consisting of ten second-order partial differential equations, given in short vector notation form by,

$$\mathbf{G}_{u,v} = k.\mathbf{T}_{u,v} \text{ where } u,v = 0, 1, 2 \text{ and } 3$$

Shock horror! This is even simpler than the unified field equations we just saw. Just in case you are still not convinced, we can safely say that all the terms in this vector equation are defined in the same way as with the unified field equations. However, notice the fact that Einstein added the electromagnetic field tensor $\tau_{u,v}$ to complete his unified field equations.

Surely it cannot be this simple. In particular, why did it take Einstein so many years to write one extra seemingly simple and innocuous-looking term to complete his unified field equations? Surely, anyone could write it in about 2 or 3 seconds. Yes, if only it was that simple. The reality is, Einstein knew it wasn't. What you have to remember is that each term contains numerous partial differential equations representing the electromagnetic field, and these have to be correctly related to the partial differential equations representing the gravitational field.

Why does this matter? It is essentially because the above equations must not violate the law of the conservation of energy. In other words, whatever energy is in the gravitational field on the left side of the equations must be equal to the energy on the right side consisting of the electromagnetic field plus a little extra gravitational energy generated in a mysterious way by the mass-energy of space from so-called uncharged matter. Einstein wanted to make sure that both sides of the equations was equal in terms of the total energy for him to see it as making logical sense and, therefore, more likely to relate to reality. This critical balancing of the equations on both sides and ensuring the total energy of the fields is properly accounted for no matter how the energy is actually transformed in reality (if at all) from one type to another (e.g., electromagnetic energy to gravitational energy and vice versa) is where the hard part lies, and potentially could take many years to get right.

But once the work is done and the unified field equations are correctly formulated and simplified into the shorthand vector notation form as we see above (thanks to Professor Tullio Levi-Civita's efforts), it helps the physicists to make a simple interpretation by reading off the unified field equations like it was a simple equation. We see this on page 220 of the 1973 book titled *The Physicist's Conception of Nature*, edited by Jagdish Mehra. Here we find the accepted interpretation given by physicists for the unified field equations,

(gravitational field or curvature) = 8π (density of mass-energy)

where 8π is part of the constant k, which has been extracted to help scientists make this simple interpretation.

In other words, the interpretation involves the concept of density of the mysterious mass-energy permeating the Universe. Sound familiar? We kind of mentioned this concept in the previous chapter. More importantly, the physicists' interpretation can be applied to both the unified field equations and the gravitational field equations no matter how complicated the partial differential equations within the vector terms may appear. In fact, if you love your straight line equation from your school days (and if so, you are a very special person), the equations can be simplified as follows:

$$y = k.x$$

where y is the strength of the gravitational field and x is the density of mass-energy. Add the electromagnetic field tensor z and the unified field equations will look essentially like so,

$$y = k(x+z)$$

Despite the addition of the term z, the interpretation given by the physicists has not undergone a radical change of any sort. That's right. As unbelievable as it may sound, there is really nothing different you have to worry about. Thus, our beloved mathematicians can do all they like to make the electromagnetic field tensor (or the other field tensors) as complicated as they like (if there is a good reason for doing so), yet the physicists' interpretation based on the concept of mass-energy density and its effect on the strength of the gravitational field is always the same.

We really cannot stress this enough. Seriously, there are no new exotic forces of nature being introduced or some other gremlin lurking behind the mathematical terms ready to pounce on the unsuspecting reader. It is all straightforward mathematics thanks to the physicists' interpretation. And, more importantly, the interpretation remains the same.

So whatever picture we must find to understand the Unified Field Theory and relate it to reality must never change at the fundamental level. All we have done is acknowledge the importance of the mass-energy density of the electromagnetic field and how it affects the strength of the gravitational field. If you like, imagine that the electromagnetic field contributes to the gravitational field, which is part of the picture we need to find.

Therefore, what the unified field equations are saying to us is that adding electromagnetic energy to the Universe (through the term z, or $\tau_{u,v}$ in the original equations) merely increases (or decreases) the energy density of this mysterious mass-energy permeating space, and this in turn increases (or decreases) the strength of the gravitational field. It is just like the interpretation we gave for the General Theory of Relativity, except instead of using acceleration of uncharged matter to affect the gravitational field, we now add the electromagnetic field and discover that it too affects this density of the mass-energy, which in turn affects the gravitational field.

Or if you want to use the mathematical vector terms from the original field equations, then we can say that this extra term known as the electromagnetic field tensor $\tau_{u,v}$ does nothing more than add extra mass (or energy) to the already pervasive mass-energy of the Universe generated by other sources (predominantly from uncharged matter) as represented by the external energy-momentum tensor $\mathbf{T}_{u,v}$, thereby increasing or decreasing the overall density of this natural mass-energy permeating space. So if the density of the mysterious mass-energy of space increases, so does the strength of the gravitational field as represented by $\mathbf{G}_{u,v}$. Similarly if the electromagnetic field tensor $\tau_{u,v}$ adds mass-energy to the Universe such that there is a cancelling out or a reduction of the density of the natural mass-energy of space, then the strength of the gravitational field is reduced as well.

Here is another way to explain it: the external energy-momentum tensor representing the mass-energy intrinsic to the entire Universe as it flows through space from known and unknown so-called "uncharged" sources as represented by $\mathbf{T}_{u,v}$ is now being *contributed to* by the electromagnetic field as represented by the electromagnetic field tensor $\tau_{u,v}$ from the simultaneous presence of known electric charge sources in the Universe.

Need independent confirmation of this interpretation by way of a quote? Perhaps this one from Tonnelat will get you going: "[$\tau_{u,v}$ is] the energy-momentum contribution of the electromagnetic field."

In other words, the word "contributing" is just another way of saying "adding". That is why we have a plus (+) sign. It is there to add extra energy to help increase the energy density of space. Turning on the electromagnetic field from charged sources has the supposed effect of adding more mass-energy to the Universe which, according to the unified field equations, can increase the strength of the gravitational field. But it is also true to say that adding more mass-energy to space by the electromagnetic field can cause destructive interference of the mass-energy when combined with whatever already exists in space. Hence it is possible that adding extra mass-energy can reduce the strength of a gravitational field.

It really is that simple.

And no, you don't have to visit a doctor to see whether your brain has had a stroke.

Finding the Right Picture

The point we are trying to make here is that the formulation of the unified field equations in the Unified Field Theory was never designed to do anything more than confirm in Einstein's own mind—based on certain thought experiments and whatever picture he had created—that the electromagnetic field had to affect the gravitational field (and possibly vice versa). Hence, the reason for the undeniable mathematical linking of the two fields in the unified field equations. The only issue scientists have with the Unified Field Theory is working out what type of electromagnetic field will affect the gravitational field. Are we referring to a static electromagnetic field, or a time-varying electromagnetic field? Whatever electromagnetic field we must use, Einstein is certain it has to affect the gravitational field as if the electromagnetic field is able to generate a gravitational field of its own and add it to the already pervasive gravitational field of the Universe from other sources.

Or to put it another way, how does the electromagnetic field affect the gravitational field in the real world? Because, if we can figure this one out, then we will know precisely what this thing in nature is to support the mathematical link, and with it the picture we are looking for and even a means of testing this link through direct experimentation.

But will this picture be enough to explain the nature of the gravitational field?

The choice of an electromagnetic field

As we have mentioned previously, scientists are happy to accept the gravitational field equations, but strangely not the unified field equations (well, until they can see how the electromagnetic field affects the gravitational field and can verify it in a real-life experiment). Yet the basic interpretation based on the "density" concept is the same. Clearly the problem is not that the Unified Field Theory is wrong or that it does not relate to reality, but the fact that scientists cannot see how the electromagnetic field affects the gravitational field and vice versa. This is the only reason why scientists have not accepted Einstein's final great theory.

So what do we know about the relationship between the two fields? More specifically, what was Einstein trying to understand just prior to creating his Unified Field Theory?

Remember from the previous chapter how Einstein was earnestly pursuing the mystery of light? Well, there really was a good reason for that. Apparently, he was deeply perplexed about a peculiar yet intrinsic property of light, known as the "particle-like" effect. This natural and inherent effect, which all scientists have come to accept, is that light has the ability to move solid matter whether that matter is charged or uncharged.

But why does light move matter?

For scientists, there is no mystery here. The particle-like effect of light can be explained in a purely Newtonian manner by accepting the concept of a photon (a packet of compressed electromagnetic energy), with a little mass from the $E=mc^2$ equation. Therefore, whatever Einstein was doing for the last 35 years of his life had to be a waste of time.

Is this true?

One does need to be extremely careful about making assumptions in relation to Einstein's work. No one knows what this mysterious matter like behavior created by light is other than for scientists to accept the behavior and have quantified it using the $E=mc^2$ equation. In other

A laser beam reflects and refracts in a convex lens. But could light be the unified field?

words, if a certain amount of electromagnetic energy hits solid matter and scientists see it move, it is equivalent to imagining a certain amount of mass from a solid object has collided with the matter. However, the equation does not explain how the mass is created or what it is precisely other than the fact that it is mixed up with the electromagnetic field; it just helps the scientists understand that light creates something reminiscent of ordinary matter.

As to whether Einstein had wasted his time with his ambitious work to unify physics, something is telling us that this can't be right. He was too confident in his work. In fact, he continued to maintain his work on the Unified Field Theory right to the end of his life. This is not a man suggesting he had failed in his work. Quite the contrary, it seems like he knew something, but he wasn't ready to tell us exactly what it was. Naughty person.

Even though Einstein has died more than 70 years ago, is there any hope for us ordinary folks to discover what he had discovered in his work?

The only way to be sure is to holistically analyze the problem Einstein had with light once more so we can see whether he was in fact wasting his time or whether he really did know he was on the right track.

Crookes radiometer

Step 1: Light is hiding something else

Our first step in the analysis must be to acknowledge the likelihood that something else is hidden in the light. Indeed, we can see that Einstein had already acknowledged this possibility. His experiment on the photoelectric effect and other observations (including the Crookes radiometer showing the movement of uncharged matter in the presence of light) told

Sir William Crookes, the inventor of the radiometer.

him there has to be something else hidden in the light to create this "particle-like" effect that influences uncharged matter. And Einstein wanted to know what this was.

Now it is our turn to find out.

Step 2: Light is an electromagnetic phenomenon

The next step in our analysis should be to acknowledge how light is a purely electromagnetic phenomenon. Certainly the famous electromagnetic genius Maxwell in the late 19th century would not deny this fact.

Now, anything electromagnetic means it should affect charged matter, right? That has always been the way electromagnetism works, or so we are told by the electromagnetic experts. Throw some electric or magnetic fields or even a combination of the two to form an electromagnetic field at matter and matter will only move if it is electrically charged.

Makes sense so far, right?

Yet when we turn our attention to the phenomenon of light, something odd occurs. It seems light can move uncharged matter. How can that be possible?

Step 3: Light moves uncharged matter

When we look at light, here we have a mysterious electric and magnetic field, oscillating (or changing its energy density over time), and made to combine in space to give the fields the remarkable ability to affect uncharged matter. Whatever these fields are, independently we know they only act on charged matter according to the laws of electromagnetism. But when we combine the fields, and the energy is oscillating, the electromagnetic field is suddenly able to move uncharged matter.

This is a verified fact. Every scientist will tell you light moves uncharged matter. If you ever need evidence of this, just look at the way the uncharged "sails" (one side is coated black and the other side is silvery) inside the Crookes radiometer spin on the tip of a fine needle when exposed to light from the environment. No battery is required to charge anything; it's all done entirely by the power of light. It is almost magical what light does.

Yet we must ask, How can this be so?

The laws of electromagnetism do not explain in any way as to why this should be the case. But because we know and can observe that light can move uncharged matter, this suggests that there is a missing factor in Maxwell's equations.

This is the crucial point. Why should light affect uncharged matter in a manner that causes movement as we see in Crookes radiometer, or any other object for that matter? There is nothing in the physics textbooks to tell us how this is achieved, especially for an uncharged object.

Is there something missing in our physics textbooks? What is really going on here?

Step 4: Light is ordinary matter

Einstein had already anticipated the problem emerging here, which is why he had to look more closely at the picture of light he was creating. At first, his important work on the photoelectric phenomenon suggested to him that light comes in packets of energy called *photons* (a name given to describe a slightly more energetic piece of oscillating energy within a certain region of space that raises the energy density to above the normal background radiation), and each photon somehow affects solid matter moving in its presence. So far, so good. The picture Einstein created seemed reasonable, and for a while, he accepted it, as all physicists do to this day. However, after noticing how these photons could move uncharged matter, he realized more work had to be done to refine the picture and help him understand exactly how photons created this "particle-like" effect.

Careful thinking on the problem would dominate Einstein's life after 1917[8]. In 1919, the results of the experiment on the solar eclipse showing that light can bend only added to the problem Einstein had with light. Why should light bend in the Sun's gravitational field? A purely electromagnetic phenomenon should never be able to do this.

---

8   Between 1915 and 1917, Einstein aimed to derive some basic solutions from his gravitational field equations.

Yet the observations are clear: light bends in the presence of a gravitational field.

Further careful thinking prevailed in the mind of the great physicist until one day, he discovered another important clue to help him formulate his new and more refined picture of light. When? We do not know for sure. Certainly it had to be well before 1922 when he published and later withdrew his incomplete paper on the Unified Field Theory.

So what exactly did he do next?

It seems Einstein reached the crucial moment in his thinking when he decided not long after 1919 to simplify the situation further by asking himself, "What is the photon?" and "How is this different from ordinary matter?"

You might think this is nothing radical. Clearly not enough to knock physicists off their seats and leave them looking like stunned mullets. And certainly you would not be compelled to jump out of your bath, run down the street naked, and yell at the top of your lungs, "Eureka! I've got it." But ask yourself the next important question regarding all matter: What does matter possess in the classical sense as we have known since the day Newton first suggested it? A gravitational field, does it not?

Now here is a rather interesting and unexpected thought. We know ordinary matter generates its own gravitational field. We see this with the Earth, a tennis ball, and anything else possessing mass. Hence, when a tennis ball is thrown through the air, starts to bend down, and eventually hits the Earth and vice versa (the Earth does move toward the tennis ball ever so slightly, but the planet is too massive to move significantly), both the tennis ball and the Earth possess a gravitational field. Since the gravitational field was theorized to affect only uncharged matter, and we assume the tennis ball and the Earth are uncharged, clearly there must be something happening in the space between the two objects where the fields interact to create this miracle of mutual attraction. Exactly how this is achieved and whether the term "gravitational field" is correct to explain it, no one knows. However, we do know that a consistent pattern is being observed between the Earth and the tennis ball (and between any other matter we can observe in the universe[9]). More specifically, the objects are drawn together. If the objects are moving and within a certain range, scientists can observe a bending of the path of each object as they move toward each other. Quite remarkable considering that this is precisely the way light bends near another object.

This brings us to a logical question: Is there any difference in the mechanism of how light and ordinary matter bend and, if close enough, eventually come together with other ordinary matter?[10] Not as far as physicists can tell at the present time, and Einstein thought so too. The path taken by light and ordinary matter when each is in range of each other is always to bend toward each other in some mysterious way, as if there is a force of attraction between them. Therefore, we have to assume the mechanism behind this observation is the gravitational effect. Based on that presumed gravitational effect (and for lack of a better description), we must call this mechanism *the force of gravitation*.

So what is the difference between light and ordinary matter?

For Einstein, he was already having an incredibly tough time differentiating between light and ordinary matter because they behave so similarly in the presence of other matter. So why separate the two? As the old saying goes, "Call a spade a spade". If we cannot see any logical and discernible difference between light and ordinary matter and the way they come together, then there is no need to create fanciful exotic particles and new forces of nature to explain the behavior. Just continue to see light as plain old ordinary matter and how it bends as entirely gravitational in nature.

So here is an interesting thought: what if light *is* ordinary matter? Because by thinking along these seemingly innocuous and simplified lines, as any good physicist would, we could see an important implication in this approach and, with it, the likely answer Einstein had been looking for to justify the work he did next. More specifically, would it not seem feasible to suggest that light, like matter, also possesses a gravitational field of its own? Because if it did, that could help explain why light moves uncharged matter, and even why light

---

9 A lowercase "universe" implies the visible universe based on what we can observe with our eyes and instruments. There is also the grander Universe containing both the visible and invisible universes. In Chapter 6, our analysis would suggest that the visible universe is probably already the Universe. As a result, this book will use the term "Universe" on a regular basis.

10 And even on itself when the density of the light or matter is raised to a high enough level.

bends in a gravitational field in a way similar to what we already know and can explain with ordinary matter.

True, such a revelation would not win Einstein a Nobel Prize for his unified field work. He would need to do just a bit more than this. Perhaps offer an explanation of the gravitational field and what it is in reality would go down well among his scientific contemporaries. But at least he got the ball rolling in the right direction by seeing a fairly obvious link between the gravitational field and the electromagnetic field.

Very clever. Here we have a rather unmistakable link, not just in mathematical terms from the field equations described above, but also in terms of what scientists have always accepted since the days of Sir Isaac Newton: the link between matter and gravitational field. As we know, a photon may be described as a packet of pure electromagnetic energy (or an electromagnetic field in mathematical terms, if you prefer such a riveting description). However, if the photon behaves like matter in every respect, that is, by virtue of its ability to move ordinary uncharged matter, why separate the electromagnetic field from the gravitational field? You need an oscillating electromagnetic field to actually generate a gravitational field of its own, which in turn gets added to the gravitational field in space generated by other sources. Then the mysterious effect of matter getting attracted to one another takes place.

Now at last, it seems we have the picture, and a real-life phenomenon to support Einstein's Unified Field Theory. Light is the unified field.

The Implications of the New Picture

Why use a fancy name?

So why do scientists complicate the situation by giving this bundle of pure electromagnetic energy the fanciful and exotic name of "the photon" when, in all probability, it is no different from any other ordinary matter? The average person on the street could easily call it a tennis ball if he wanted to. Of course, it would be a rather small and very fast tennis ball. Or why not call it the "fuzz ball"? Either way, it doesn't matter. It is just a name we give to this bundle of energy moving at the speed of light. You can call it whatever you like. If you want to call it Donald Duck, that is perfectly fine. In quantum theory, light is not seen as real, but rather a fictitious bundle of energy that may or may not be traveling through space. The only time this picture changes and for us to acknowledge the energy exists and is a real phenomenon is when light collides with solid matter and that is when we have to call it something else. So prior to the collision, you can call the photon a fictitious character, such as Donald Duck. It is not an unreasonable name to give to this energy. However, it might be worth calling it something else to suggest it is real as soon as it hits ordinary matter.[11]

Might as well call it ordinary matter.

At the end of the day, it is totally up to you how you want to call this energy. Just remember to see light (or the photon) for what it is: light is ordinary matter.

How does light create a gravitational field?

Good question. You have come to a major sticking point in this whole idea. Understanding how light creates a gravitational field is basically the same problem of how light creates ordinary matter. No one really knows how light does this. Mass and the gravitational field have to be seen as natural and inherent properties of light. In other words, we must accept these properties. As terrible as this may sound to the scientists who prefer to question everything, sometimes we do have to be like some religious leaders in learning to accept certain things. Nevertheless, there are some interesting observations that seem to make this gravitational field in the light seem consistent with everything we know.

In Chapter 2, we mentioned how ordinary matter creates a gravitational field of its own. According to the General Theory of Relativity, this is because ordinary matter is accelerating. Thus, simple forms of acceleration, such as vibrating and spinning by the atoms and subatomic particles making up matter as well as the entire object itself,

---

11   This reveals a kind of paradoxical behavior in the way the electromagnetic energy moves around in space: either it exists or it doesn't. So long as the energy does not interact with solid matter, there is always a paradox in this energy. Speaking of a paradox, we will have more to say about this in Chapter 8.

can create this gravitational field. As an example, we know that the atoms in the crystalline structure making up ordinary matter are never at a complete rest. They are being hit by radiation in the environment, as well as the accelerating motion of the electrons, protons and neutrons making up the atoms. While the atoms absorb the energy in the radiation, they can emit energy as well, which in turn causes the atoms to recoil and move in the opposite direction, thereby maintaining its acceleration (i.e. vibrating motion). And while the atoms accelerate, somehow this emission of energy carries with it a gravitational field into space to influence other matter. Likewise, if the atoms could somehow remain motionless in space, the entire object vibrating or spinning can also absorb and emit energy from the environment to help create the gravitational field.

In the case of the oscillating electromagnetic field, the generation of a gravitational field would suggest that the electromagnetic field must be accelerating.

How?

It must have something to do with the oscillating behavior of the energy. When the electromagnetic energy vibrates like the plucked string on a guitar, the gravitational field is created. But since the speed of light never slows down the further it travels through space, the electromagnetic field, when oscillating, must be self-accelerating to maintain this maximum speed. How it actually does this is not entirely clear. The closest analogy we have at the moment can be found in classical electrodynamics when we look at the behavior of an electron emitting radiation. According to the solution to the Abraham-Lorentz formula for this situation, the electron self-accelerates as soon as it emits the smallest amount of radiation from its charged surface in one direction. But the radiation does not disappear into space. The energy in the radiation is somehow being brought back to the charge in a kind of "gravitational attraction" manner by the accelerating charge to be re-used for the next radiation emission. Indeed, the literal interpretation of the solution, and one that does not violate the laws of energy conservation, is that the accelerating charge is recycling the emitted energy, and the recoiling force on the charge by the constantly emitting and re-absorbing of the radiation is driving the charge to accelerate faster and faster.

Perhaps a similar mechanism might be taking place within radiation itself in order to explain how it is able to maintain maximum speed? In other words, could the mass-like property of light be acting as a kind of springboard from which the energy can push against, but its gravitational field is helping to keep the energy confined to itself while the entire photon is made to self-accelerate in response to this energy emission?

Whatever is happening, we can imagine light as (self-)accelerating to be consistent with the General Theory of Relativity when creating a gravitational field.

Matter can travel at the speed of light

As another interesting implication to emerge from this new picture of light and its gravitational field is the maximum speed of ordinary matter. Specifically, we can now say with reasonable certainty that ordinary matter can travel at the speed of light.[12] Before, scientists stated that matter can never attain the speed of light. Well, not anymore. According to Einstein's Unified Field Theory, ordinary matter should travel at the speed of light.

Want matter to go faster than the currently accepted speed of light? No problem. Do something to lower the density of this mysterious mass-energy of space and you will always be able to exceed the standard speed of light (i.e., 300,000km/s).

However, if you are talking about the same density, then you have to say matter can never *exceed* the speed of light. It can equal it, but it can never exceed it.

The Picture is Clear

Whatever we should call this bundle of compressed electromagnetic energy (a photon, a nano-sized tennis ball, the fuzz ball, Donald Duck etc.) and what makes the oscillating behavior such a difference in creating this effect of ordinary matter and the gravitational field to help this energy move other energy and matter, we do have to admit that

---

12   The speed of light is controlled by the mysterious mass-energy density permeating all of space. Thus, if you reduce this density, ordinary matter can appear to travel faster than the speed of light. However, you have to remember that light travels faster in the same density as well. Therefore, in reality, matter never exceeds the speed of light.

this is an amazing coincidence. It would appear that the link so often overlooked by scientists and suggested by Einstein as possibly existing through the unified field equations is actually apparent in the rather common phenomenon we call light.

Did seeing light as ordinary matter cause Einstein to combine the electromagnetic and gravitational fields into one mathematical superstructure?

Well, one thing seems clear: it would be silly to leave out the gravitational field from light if light is behaving in every respect like any other matter. Einstein was quite happy to use the famous $E=mc^2$ equation to link any form of energy with matter at the moment it collides with other matter. So if light is pure energy and is able to act like matter in the way it moves uncharged matter with such incredible ease, this equation would therefore suggest that light has a gravitational field and acts like a solid piece of ordinary matter under Newton's law of gravitation.

So why not make the link between the electromagnetic and gravitational fields in the case of light?

It just makes sense.

And with this link came into existence Einstein's final masterpiece known as the Unified Field Theory.

What is the Gravitational Field?

Despite finding a remarkably simple interpretation and having light as a logical, real-life picture that fully justifies Einstein's decision to develop his Unified Field Theory, there is one more kink in the armor behind our unified field idea. Apart from not knowing how the electromagnetic field generates a gravitational field and ordinary matter, we still have not identified the gravitational field. Remember from our previous chapter: the whole point of doing this work was to figure out the nature of this mysterious gravitational field. What is it exactly? And what precisely generates it? So far, we know the electromagnetic field is one source for generating the gravitational field. But if we took away the electromagnetic field from the Universe and returned to the original gravitational field equations, we are still left with the uncharged matter somehow able to generate a gravitational field and so influence the density of this mysterious mass-energy of space. How does uncharged matter create the gravitational field?

What complicates this situation is the Unified field Theory claiming the electromagnetic field, when oscillating, generates its own gravitational field. The electromagnetic field is presumably not relying on subatomic particles or solid matter to create it. It is not even suggesting that light is carrying some of this gravitational energy from solid matter. Rather, the theory is suggesting that light on its own can generate the gravitational field.

But how does light do this? Or is it the accelerating charge that generates both the electromagnetic field and the gravitational field at the same time?

One observation to have physicists in total agreement is the fact that anything that is charged always carries with it a mass. Whether it is a negatively charged electron or a positively charged proton, these subatomic particles always have mass. So is it the mass that is the deciding factor in determining the source of the gravitational field, and the electromagnetic field could be merely carrying this gravitational energy into space so as to give the impression that it is generating the gravitational field?

This brings us to a more fundamental question, How much of a contribution is the oscillating electromagnetic field making to the generation of a gravitational field? Because if there is any way the gravitational field can be generated independent of the electromagnetic field and electric charge, physicists can focus more closely on the mass to see how it might be achieving this miraculous feat with the gravitational field.

To understand the true nature of the gravitational field and where it is coming from, we have to look more closely at the external energy-momentum tensor $T_{u,v}$. This is the term common to both the Unified Field Theory and the General Theory of Relativity. It represents the mass-energy of space generated by uncharged matter in the Universe and the thing that constitutes solid uncharged matter (i.e., presumably made of the same mass-energy). It means that there is something inside uncharged matter that is creating this extra mass-energy and, with it, the gravitational field. Additionally, it is emitting this gravitational energy to its surroundings to affect the overall energy density of the mass-energy of space.

If we can identify the source of this extra mass-energy coming out of uncharged matter, not only can we solve this mystery, but we can potentially understand the nature of mass-energy itself. Indeed, much talk has been made in Chapter 2 of this mysterious mass-energy permeating the Universe as represented by the external energy-momentum tensor $\mathbf{T}_{u,v}$. Many readers must be wondering, what exactly is this mass-energy?

One thing is certain, whatever this mass-energy is, it can be destructively or constructively interfered by the addition of electromagnetic energy, causing the mass-energy density to decrease or increase, respectively, which in turn affects the strength of the gravitational field. For any electromagnetic field to achieve a canceling or amplification of the mass-energy density, it would suggest that the mysterious mass-energy is probably electromagnetic in nature. Could this be nothing more than the natural background radiation flowing through space?

To find out if this is true, we begin by asking ourselves:

> Are all the sources of the electromagnetic field in the Universe properly taken into account when determining the source of the gravitational field?

In the dreaded language of mathematics that thrills readers to no end, all we are saying is, What is the external energy-momentum tensor $\mathbf{T}_{u,v}$ in reality?

Surely, this would have to be one of the great problems Einstein had to solve toward the end of his life. For if additional electric charges in the Universe not accounted for are possibly lurking within this term, they could potentially be contributing to the creation of not only the electromagnetic field but also the gravitational field. To be sure this is not the case, Einstein would have attempted to identify all sources of the electromagnetic field and imagine these sources disappearing from the entire Universe. Then whatever is left would represent (and act as the source of) the mysterious mass-energy of the external energy-momentum tensor $\mathbf{T}_{u,v}$. That is when we would know if the gravitational field were created not by electromagnetic sources but rather from more exotic sources lying inside uncharged matter, and potentially within the subatomic particles of the electron, proton and neutron. And if that should be the case, then the unified field equations would be brought back to the original gravitational field equations of the General Theory of Relativity (representing a Universe completely devoid of charges and the electromagnetic field), thereby proving once and for all that something else of non-electromagnetic origin must be controlling the gravitational field. But if not, and the mass-energy represented by $T_{u,v}$ is entirely electromagnetic in nature, then the removal of all charges in the Universe should cause the unified field equations to collapse to,

$$G_{u,v} = 0$$

which means that no gravitational field exists, thereby proving that the electromagnetic field is not independent of the gravitational field. That would mean that light must be creating the gravitational field, and nothing else can produce it.

For us to identify all sources of electric charge present in the Universe, we must conduct one simple thought experiment. That experiment concerns all uncharged matter in the Universe—namely, is it really uncharged?

The truth about uncharged objects

For all hypothetical purposes, let us imagine a supposedly uncharged object sitting in an otherwise empty Universe. The next thing we do is grab the most powerful microscope you could ever imagine and look at the surface of this uncharged object. What we should see are numerous vibrating atoms sitting on the surface, held together in a rigid crystalline structure by the electrons in the outermost orbits of each atom and moving between the atoms to form chemical bonds; the rest of the electrons in the innermost orbits going around each atomic nucleus are there to help neutralize the positive charge of the protons in the atom's nuclei.

Now suppose a super-microfine electrode tip of our imagination could be acquired. We connect it to a super-sensitive meter designed to measure electric charge at any instant. The electrode is then placed over a tiny area directly above a single atom. What do you think would happen next?

As the electrons accelerate rapidly in their various discrete orbits around the atom and also in the chemical bonds, we would notice how, at certain times, an excessive number of electrons would appear for a fleeting moment on the surface of the object closest to the electrode tip. Because electrons are negatively-charged particles, these extra electrons would register a negative charge on the meter.

We then wait for a fraction of a second and measure again.

As the electrons move around to another position, there would be moments when the nucleus of the atom, which contains positively-charged protons and the uncharged neutrons, is marginally exposed. When this happens, the meter would detect a slightly positive charge over the same area.

We keep measuring the charge over this tiny area. As we do, we notice how the electrode will pick up and register on the meter a constant variation in the electric charge as it swings from negative to positive and back again.

But according to our classical laws of electromagnetism, we are told that, by definition, a constantly changing electric charge over time as displayed by the meter implies the presence of an electromagnetic field emanating into space from this tiny area directly above one atom. Unless the laws are completely wrong, we would have to say that the uncharged object is in fact charged and is generating time-variable electromagnetic fields of its own all around the entire surface of the object.

Here is a rather interesting discovery. Our simple thought-experiment suggests that a so-called uncharged object is constantly charged and spuriously emitting radiation into space. It is just that when we touch the matter with our fingers or the tip of an electrode, the large area of these surfaces, the speed at which the charge swings between negative and positive, and the effect of literally so many atoms doing the exact same thing while generating different charges over time, yields an average result on the meter that effectively makes us think that the object is uncharged at all times. Yet our thought-experiment patently shows this to be untrue. Uncharged matter is actually always charged.

There is direct proof for this claim. Wherever we look, solid and supposedly uncharged matter is visible to our eyes. For something to be visible, this is only possible if the matter is emitting radiation to reach our eyes. Far from cancelling out the energy from the great multitude of atoms emitting radiation of their own, the fact that we can observe anything in this Universe proves that uncharged matter is charged and spuriously emitting radiation in all directions.

This is now telling us that there are additional electric charges unaccounted for, and these accelerating charges (that is, charges that are spinning and orbiting the nuclei of atoms) are likely to contribute to the creation of the gravitational field through the radiation they emit. It isn't just the electrons. The protons are contributing to the creation of an electromagnetic field. In fact, the nucleus is spinning and emitting what is called *synchrotron radiation* and used to hold the electrons in specific orbits. If we are to find the source of the gravitational field and with it, hopefully, the true nature of the natural mass-energy permeating the Universe, all these charges making up the atoms of ostensibly uncharged matter must be taken into account and placed into the electromagnetic field tensor $\tau_{u,v}$. Then the tensor must be eliminated from the unified field equations, which is equivalent to turning off all electric charges in the Universe using our vivid imagination.

So what is left in the Universe when we do this?

After eliminating the electrons and protons in the Universe, we are left with the neutrons. And as scientists would tell us, these are truly uncharged particles, right? So the origin of the gravitational field must be this particle, and the electromagnetic field cannot be generating the gravitational field.

Therefore, Einstein must have failed in his ambitious quest to understand the mystery of gravity. The Unified Field Theory was really a waste of time.

Well, hold your horses. Do we really know the neutron is uncharged at every single moment in time? Given what we now know about uncharged matter, it may be wise to consider whether the neutron is charged as well. Why? For the very reason that a neutron is composed of an electron and a proton, as scientists have testified on numerous occasions after smashing a neutron inside a particle accelerator. But you might be thinking that surely the electron and proton have combined inside the neutron to completely cancel out the charge. Are you sure about this? In theory, the electron and proton might act as if they are combined. Experiments might suggest no charge to support this view but only because the instruments we use are not sensitive enough at every instant in time to measure the charge of the neutron. We have to realize that what the instruments are doing is taking an average result by sampling the environment they are trying to measure over a given time frame, displaying it on a screen, and giving the perception to the scientist that no charge exists, all because the instruments are not perfect. In other words, the instruments cannot sample the charge quickly enough to know exactly what is happening to the neutron. If this is true, there is every chance that the electron and proton exist in their separate charged entities spinning around each other to give the impression to an outside observer of a single uncharged particle called the neutron.

To give an analogy, consider a boy and girl holding hands and spinning around each other. The boy's force keeps the girl standing, and her force keeps him standing. As long as neither lets go, the pair will continue to spin in place. But if either the boy or the girl weakens the force in any way and suddenly lets go, the pairing will separate, and the pair is destroyed. We suspect that this is probably the real situation taking place inside the neutron: the boy is the proton, the girl is the electron, and they are doing their famous electromagnetic dance with each other. But should there be any disruption to the dancing, the neutron will be destroyed.

Thus, the only thing suggesting that these particles are probably not combined properly in the neutron is observing how easily the electron can leave the neutron in an atom's nucleus, leaving its "partner proton" behind, then join a different proton to create a new neutron. Because it seems unlikely for an electron to want to leave one proton inside a neutron to join another proton, scientists assume the force responsible for pulling out the electron is not electromagnetic but something more exotic called the *weak nuclear force*. But has this force always been electromagnetic, and scientists just haven't realized it? We will have more to say about this in the conclusion. For now, let us continue with our present discussion.

Setting aside what this weak nuclear force might be, it might be presumptuous to assume that the neutron is uncharged until our instruments are sensitive enough to prove that conclusively. Until then, to be sure that we have captured all the electric charges, the electron and proton of the neutron must be considered when eliminating charges from the real Universe. Otherwise, there is a possibility that the neutron is somehow leaking its own electromagnetic energy of its own into the environment in a way we cannot detect, making us think it is uncharged when in fact, at the subatomic level, it could well be constantly charged (i.e., swinging from negative to positive charge very quickly).

So what's left? Well, a neutron can decay into an electron, a proton, and some remaining energy called an *anti-neutrino* originally designed to hold the two fundamental particles close together and yet still keep them apart inside the neutron. Sure, sometimes there is a delay in the decay process in which the electron does not appear once the proton emerges. In other words, the electron can be bonded to some of that remaining energy to form a W-boson, but quickly decays into the electron and anti-neutrino.

Therefore, the gravitational field must be coming from this anti-neutrino, right? Highly unlikely. Like neutrinos, these are thought to have zero mass and travel at the speed of light, but others have claimed they can have a small mass value but cannot travel at the speed of light. If it is the former description, it is highly reminiscent of the properties of radiation, as if it could be electromagnetic in character. This is a bit like the gravitational field in the way it moves at the speed of light, except we are trying to find out here if it is electromagnetic in nature (i.e., created by radiation). But assuming that no energy (or mass) can move at the speed of light except radiation, a neutrino and an anti-neutrino that can travel at the speed of light must be radiation. If so, it seems we are returning to the idea that an electromagnetic field is controlling the gravitational field.

However, if the neutrino does contain some mass and, therefore, cannot travel at the speed of light, it makes an even less likely candidate to control the gravitational field. The problem is that the neutrino (and, by all indications, the anti-neutrino) is especially hard to detect and study because this energy packet is so dense and small—and so small and fast that it rarely interacts with other particles of matter. This is the opposite of what you want when explaining the behavior of a gravitational field because the gravitational field must interact with all matter to make it move. To give you an idea of just how rare the interaction between neutrinos and matter is, Katlyn Edwards said the following:

> "In fact, a neutrino would have to pass through several thousand light years of solid lead before it would have a 50-50 chance of being absorbed."[13]

If that is not enough, John F. Beacom and Nicole F. Bell from the University of Melbourne acknowledged the possibility that neutrinos could "decay into truly invisible particles"[14]. For all we know, these particles may be nothing more than photons merging with the rest of space. Gravitational fields, on the other hand, do not decay. They exist indefinitely.

So what remains? With neutrinos and anti-neutrinos not in the required numbers to account for the gravitational field or they appear indistinguishable from radiation, we are not left with many options. If there is any remaining energy in the universe to consider, it must be related to the strong and weak nuclear forces. These are the forces that somehow bind an electron to a neutron and keeps the protons together in the nucleus. However, they act over distances far too short to account for the influence of the gravitational field over planetary and stellar distances.

With no other sources of gravitational fields to consider and all charges eliminated, including those residing in the neutron, we must conclude that the only thing capable of controlling the gravitational field has to be the electromagnetic field. The electromagnetic field is the only force of nature physicists know of that has the required range of influence on solid matter (which is infinite, just like the gravitational field), travels at the same speed (both the gravitational and the electromagnetic fields travel at the speed of light), and can influence solid matter in a remarkably similar way during a collision.

But is the oscillating electromagnetic field actually the gravitational field?

---

13   https://www.physicsforums.com/threads/why-neutrinos-rarely-interact-with-matter.434892/
14   Beacom & Bell 2002.

Well, after hypothetically removing all known sources of the electric charge within the neutron throughout the Universe to stop the electromagnetic field from being generated, we discover something remarkable: all matter (and energy) in the Universe disappears. That's right. Remove all the electrons and protons from absolutely everything, including the neutron, and let the remaining energy that binds the two charges into a neutron decay to pure electromagnetic radiation until there is nothing left, and the electromagnetic field tensor $\tau_{u,v}$ drops to zero. In other words, the electromagnetic field disappears as there is no charge available to generate this field. More amazingly, removing all charges from the Universe also has the effect of removing all solid matter from the Universe. This means the external energy-momentum tensor $T_{u,v}$ drops to zero, and that means the mass-energy of space generated by solid matter disappears too. Therefore, this has the effect of eliminating the gravitational field altogether. The very thing we are trying to understand has suddenly disappeared from our hypothetical Universe in an attempt to separate the electromagnetic field from the gravitational field. This simple thought experiment strongly implies that in order to have a gravitational field permeating the Universe, it must always come with an oscillating electromagnetic field. You cannot separate the two fields under any circumstances. If one field exists, so too does the other.

If what has been said so far is true, then the sources of the mysterious mass-energy permeating the Universe as represented by the tensor $T_{u,v}$ must come from charged matter. The charged matter contained in so-called uncharged matter must be combined with the electromagnetic field tensor. And that means $T_{u,v}$ must merge with the electromagnetic field tensor $\tau_{u,v}$. Therefore, the unified field equations only requires one universal electromagnetic field tensor, and that means the equations get simplified to:

$$G_{u,v} = k.\tau_{u,v} \text{ where } u,v = 0, 1, 2 \text{ and } 3$$

As for the mysterious mass-energy of space responsible for affecting the gravitational field through a variation in its density, this is nothing more than radiation. Radiation, or the oscillating electromagnetic field, is creating the effect of gravity and universal gravitation, depending on its density. And while the energy density oscillates, it will generate the mass-like property (and the gravitational field) needed to push solid matter. Now, if there is a mechanism to show this pushing action can help to bring matter together in some natural way, it could well be the thing that will explain what scientists have previously called the "gravitational effect".

Is there a way for matter to naturally clump together through the pushing action of radiation on matter? We will learn more about that in the next chapter.

Is Matter Purely Electromagnetic?

Assuming the neutron is charged, this disappearance of all matter in the Universe, including radiation, must have seriously disturbed Einstein because it would mean that the thing responsible for creating the gravitational field cannot be separated from the electric charges composing all charged and supposedly uncharged matter. This is clearly telling us the electromagnetic field is the source of the gravitational field. And it is all because the matter we see around us, charged or otherwise, is composed entirely of these two fundamental atomic charged particles known as the electron and proton. As E.R. "Boblock" Le Clear wrote on his web site:

> "There are only two types of stable charged particles, the proton (sink) and the electron (source). All other stable matter is composed of protons and electrons including the neutron."[15]

Boblock further added that mass is defined as "being electromagnetic in character..." and specified that "mass cannot exist without charge."

It is interesting to see Boblock giving his firm view that charge is of only two types: electrons and protons. But is there meant to be an exotic particle capable of surviving in a stable and independent state like the electron and proton can and which would thus be capable of being the basis for producing the gravitational field? Well, possibly. But

---

15   Boblock's email address as of June 1999 was boblock@montego.com. Miodrag Malovic now hosts his ideas on the Unified Field Theory at http://www.beotel.yu/~mmalovic/boblock/.

you would think that after all this time, the scientists would have found this exotic particle in large quantities acting in a very stable state.

To date all the exotic particles discovered by scientists are in limited quantities and have incredibly short lifespans in the order of 10 nanoseconds (0.00000001 seconds), making them very unlikely candidates. If a truly stable exotic particle does exist, we would have detected it almost as quickly as we have found the proton, electron, and neutron.[16]

So far, the evidence is lacking.

The Electromagnetic Universe

It would seem that if you are one of those physicists wanting to stick to the old school of thought that gravity and light are not related, you now face the unpalatable possibility that the electromagnetic field is the source of the gravitational field. This means that the external energy-momentum tensor representing the mysterious natural mass-energy of space has to be the natural background radiation. Yet we are no closer to understanding what the gravitational field is. We may realize that the gravitational field is probably mixed up in the oscillating electromagnetic field as if the latter field is acting as the source for the gravitational field, but we are no closer to reaching the ultimate holy grail of physics.

Can we explain the gravitational field?

Well, what if Einstein had decided at some point late in his life to ignore the existence of the gravitational field? Why have it at all if we do not know what it is?

An interesting idea. Clearly, the only thing that would be left is the electromagnetic field. And it would leave physicists with just one ultimate question: what is the electromagnetic field? Furthermore, this idea would force all physicists to consider the Universe from a purely electromagnetic perspective. In other words, the only force of nature in the Universe is electromagnetic and produced by radiation and the way to unify all of physics is under the umbrella of electromagnetism. No exotic forces of nature, such as the weak and strong nuclear forces, to

---

16   Even the Higgs boson, or Higgs particle as it is known, used to explain why the weak nuclear force acts over a much shorter range than the electromagnetic field and why some fundamental particles have mass, is not able to last long enough to control the gravitational field with its estimated lifespan of 1×10^{-33} seconds.

think about. There is no gravitational field to consider. There is only the electromagnetic field, when oscillating, that creates the Universe and helps us to see everything, including how matter clumps together naturally.

This is a very subtle paradigm shift in thinking. It involves jumping off the old gravitational bandwagon and boarding the supertrain of electromagnetism to see the kind of Universe we could actually be living in.

True, we do not know whether Einstein made this bold attempt to ignore the gravitational field, but it would seem like a reasonable step to take. Seriously, why would anyone want to solve two big problems when they could be reduced to just one? If Einstein had been thinking along these lines, the result would be a purely electromagnetic Universe where all the forces of nature are purely electromagnetic. Nice and simple. Now that is an amazing thought. Maybe this explains why Einstein felt the world was not ready for this next major revelation and decided he would keep this part quiet.

Of course, this is all speculation at this point.

However, certainly the biggest advantage of this approach is how we can easily reduce our problems to just one question: what is the electromagnetic field?

The only thing is, understanding the electromagnetic field in its absolute fundamental form and not rely on mathematical descriptions to describe it remains the biggest mystery for physics. It is probably akin to asking a religious person, what is God? Except you are asking a scientist to ask the exact same question for light. What is light? No one really knows. We only have certain properties, such as the ability to move ordinary matter But what it is in reality is something no scientist can explain.

Did Einstein find a way to break through this ultimate barrier of physics?

Evidence Supporting the New Interpretation

View of Dr Leopold Infeld

Suppose Einstein didn't realize that the electromagnetic field is the gravitational field. Are we on the right track here in following Einstein's likely reasoning for light when he explained the "paradoxes arising out of light" as due to the presence of a gravitational field? If we are to maintain this "gravitational field" concept in physics, there is confirmation for the idea in statements made by Polish physicist Dr. Leopold Infeld (1898–1968), an assistant[17] to Einstein in the 1930s and a professor of applied mathematics at the University of Toronto, Canada.

In 1950, Infeld wrote provocatively about his understanding of the Unified Field Theory as follows:

> "The gravitational field is influenced not only by the moving [accelerated] gravitational masses but also by the electromagnetic field. Thus the sources of a gravitational field lie in moving [accelerated] masses, in moving [accelerated] charges and in the electromagnetic field."[18]

Because light is effectively an electromagnetic field, this statement would appear to hold true for all forms of radiation whether infrared, ultraviolet or gamma rays (assuming this statement refers to oscillating fields).

He further goes on to say that:

> "A pure gravitational field can exist without an electromagnetic field. But a pure electromagnetic field cannot exist without a gravitational field."[19]

The first part of his sentence is interesting because it suggests that Infeld was not aware of the charges present in uncharged matter or does not see a possible contribution from these charges to the generation of the electromagnetic field that might serve as the source of the gravitational field. He is looking at the unified field equations in

---

17   Bernstein 1991, p.55.
18   Infeld 1950, p.115.
19   Infeld 1950, p.115.

a literal sense and sees the energy-momentum tensor $\mathbf{T}_{u,v}$ as a necessary term because he believed uncharged matter exists in the Universe. He never thought of the possibility that uncharged matter is always charged. So although matter is seen as uncharged in Infeld's mind, we can understand why he believed the gravitational field can exist without the electromagnetic field.

At any rate, Infeld does firmly acknowledge that a "pure electromagnetic field cannot exist without a gravitational field". From our previous analysis, we can understand this statement to be correct. The only problem we have with Infeld's statement is how he does not explain what type of electromagnetic field creates a gravitational field. He just fell short on this aspect. So all we can say is that he may be indirectly supporting the concept of light having a gravitational field of its own.

Is there another quote to give greater credence to this concept?

As further evidence for this concept, the *McGraw-Hill Encyclopedia of Science and Technology* states:

> "...the electromagnetic field contains energy and is thus the source of some of the curvature of the space."[20]

But because the curvature of space-time actually implies the presence of a gravitational field according to the General Theory of Relativity, the only way this statement can hold true is if the electromagnetic field is somehow creating its own gravitational field to help it affect the curvature of space-time. Therefore the electromagnetic field must have a gravitational field. Again, we are not sure if the electromagnetic field has to be oscillating to achieve this "bending effect".

However, in the article titled "The Connection between Gravity and Electricity", published in *English Mechanics* on May 8, 1936, W. D. Verschoyle writes:

> "...Einstein deduced mathematically that light should be deflected in a strong gravitational field, and careful observation during the total solar eclipse of 1919 showed this definition to be a fact....Since gravity affects some forms of

---

20   *McGraw-Hill Encyclopedia of Science and Technology* 1992, Seventh Edition, Volume 15, p.294.

radiation, some form of powerful radiation will almost certainly be found to affect gravity."[21]

As this quote tells us, it seems someone is convinced that an oscillating electromagnetic field (or radiation) can affect gravity. For this to occur must imply the presence of a gravitational field in the radiation. Unfortunately, what is in the radiation that would affect gravity (i.e., whether it is the gravitational field) is not specified in the article. The author merely acknowledges the reality of this observation but does not go further to explain why light bends in a gravitational field. It just is. Nonetheless, the author does feel confident that radiation should be able to influence gravity.

Here we have the clearest indications from the literature that an oscillating electromagnetic field must affect the gravitational field by creating its own gravitational field to help add to what is already present in the Universe.

The Philadelphia experiment

As a result of this insight about radiation, it is quite possible that someone other than Einstein may have stumbled on this idea and pursued it through a secret experiment.

In the late 1970s, researcher William Moore interviewed a man by the name of Dr. J. Manson Valentine, an oceanographer, zoologist, and archaeologist. During the interview, Valentine made the following statement concerning the 1959 conversation he had with a close friend, Dr. Morris Ketchum Jessup (a lecturer of astronomy and mathematics at Drake University in Iowa, USA, in the 1950s), just before Jessup's untimely death that same year:

> "...In practice, it [the secret United States Navy project on invisibility allegedly conducted in 1943] concerns electric and magnetic fields as follows: An electric field created in a coil induces a magnetic field at right angles to the first; each of these fields represent one plane of space. But since there are three planes of space, there must be a third field, perhaps a gravitational one. By hooking up electromagnetic generators

---

21   Verschoyle 1936, pp.78-79.

so as to produce a magnetic pulse, it might be possible to produce this third field....Jessup told me that he thought that the U.S. Navy had inadvertently stumbled on this."[22]

The reader should be reminded that a pulsing magnetic field is equivalent to saying there is a time-varying electromagnetic field (i.e., radiation). So if Dr. Infeld's statement regarding electromagnetic fields remains true, then this pulsing electromagnetic field must have a gravitational field.

According to Dr. Jessup, it is the gravitational field generated by the pulsing magnetic field that affects the path taken by light around an object (in this case, a U.S. Navy ship), allowing observers to see what is behind the object and so render the object invisible.

Can science create invisibility?

Looking at the scientific literature, we find evidence of researchers at Duke University, North Carolina, and at Imperial College London attempting to achieve invisibility.

An article published on the *Science Express* web site (the online version of the journal *Science*) on May 25, 2006, claims that researchers have the mathematical equations to prove that the invisibility phenomenon can occur in reality and are developing a cloaking material using metamaterials to achieve this very aim. As David R. Smith, professor of electrical and computer engineering at Duke's Pratt School of Engineering, put it:

> "Light would flow around an object hidden in the cloak just as water in a river flows virtually undisturbed around a smooth rock."[23]

In other words, instead of bending light as we might expect using the Unified Field concept of increasing the strength of the gravitational field of light through a raising of the energy density of the time-varying electromagnetic field, the researchers are looking into nanotechnology to create a range of exotic and rigid metamaterials with the ability to deflect the light around the object without scattering the light waves.

---

22   Berlitz & Moore 1979, pp.131-132. Earliest published version of the quote.
23   Dr. Smith's quote obtained from http://www.dukenews.duke.edu/2006/05/cloaking.html. NOTE: *Science Express* web site is at http://www.scienceexpress.com.

For clarity, metamaterials are defined as composite materials capable of exerting desired effects on electromagnetic fields. These materials were first pioneered in 2001, opening up a wealth of new scientific possibilities including, naturally enough, invisibility.

As Smith said:

> "The cloak would act like you've opened up a hole in space. All light or other electromagnetic waves are swept around the area, guided by the meta-material to emerge on the other side as if they had passed through an empty volume of space."[24]

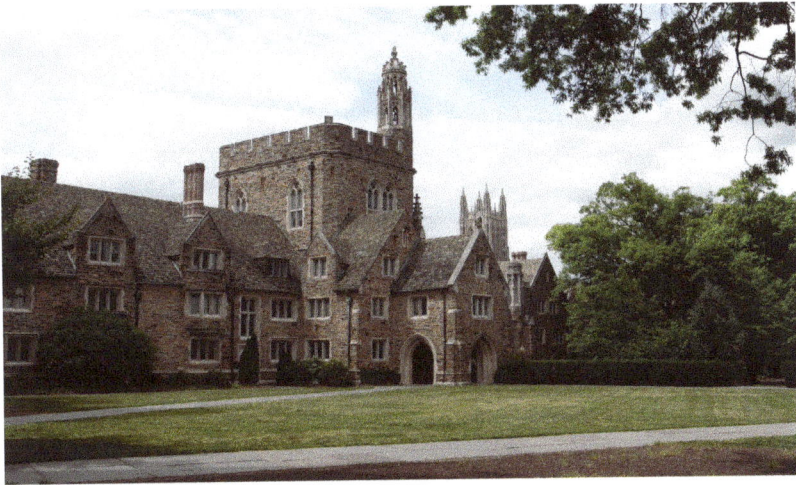

Duke University where the latest research into invisibility using metamaterials is being carried out.

Dr. David Schurig, an associate researcher in electrical and computer engineering of Duke's Pratt School, explained the idea in these terms:

> "You could think of space as a woven cloth, and if you wanted to make a hole in space to hide something, you could imagine sticking a pointed object in between the threads and distorting them and making a hole. The electromagnetic waves, or light, are constrained to move along the threads. So the light waves come along and spread around the hole and then return to their original configuration..."

24   http://www.dukenews.duke.edu/2006/05/cloaking.html.

If you had come up with it 10 years ago, you would have looked at the materials specification and said, 'I can never build this'. And now, because of this meta-material research field that has grown, this becomes more interesting, because we can actually build the things this specifies."[25]

To obtain further details, David Schurig and Sir John Pendry have reported their findings to *Science Express*.

In March 2007, the idea of deflecting light went one step further when a group of engineers at Purdue University in Indiana realized it might be possible to use ordinary metals shaped into tiny needles to do all of this light-bending work. The process would involve fixing large numbers of needles inside a cone-shaped object such that anything placed inside the cone is rendered invisible to the naked eye. As Vladimir Shalaev, professor of electrical and computer engineering at Purdue University, said:

"It looks pretty much like fiction, I do realize, but it's completely in agreement with the laws of physics.

Ideally, if we make it real it would work exactly like Harry Potter's invisibility cloak. It's not going to be heavy because there's going to be very little metal in it."[26]

Further details about the cone of invisibility idea can be found in the April 2007 edition of *Nature Photonics*.

Despite this interesting work, we shouldn't get our hopes up too high for metamaterials or the "cone of invisibility" to render invisible anything as large as a ship any time soon as was allegedly the case in the secret U.S. Navy experiment in 1943. Modern researchers are only interested in the feasibility of the idea and seeing whether a metamaterial or a bunch of metal needles can be made to deflect the light around it.

But what about using the oscillating electromagnetic fields to achieve the same thing?

Smith stated:

25   http://www.heraldsun.com/durham/4-738683.html.
26   *The Canberra Times*. "Invisible invention". April 7, 2007, p.19 (Engineers at Purdue University and quote by Professor Vladimir Shalaev).

"The theory [behind invisibility] itself is simple. It's nothing that couldn't have been done 50 or even 100 years ago."

So, did the U.S. Navy succeed in rendering a much larger object invisible through the unified field concept? No one can say for sure whether this happened. The only evidence we have are letters from a mysterious man named Carlos Allende claiming to have witnessed the event, and an official photograph taken of two U.S. Navy officers in Einstein's home in 1943. The U.S. Navy has allegedly lost all records on the *Eldridge*, the ship that was claimed to be involved in the experiment during its sea trials many decades ago.

The only known photograph of two U.S. Navy Officers conferring with Einstein in his home in Princeton on July 24, 1943, believed to be a memento for the U.S. Navy after discovering something about the Unified Field Theory. Prior to this photo opportunity, Einstein had been engaged in research for the U.S. Navy's Bureau of Ordnance as a consultant on high explosives on May 31, 1943. Seated is Captain Geoffrey E. Sage, USN, commanding officer, Naval Training Station, Princeton, and standing is Lieutenant Commander Frederick L. Douthit, USNR, executive officer, Naval Training Station, Princeton.

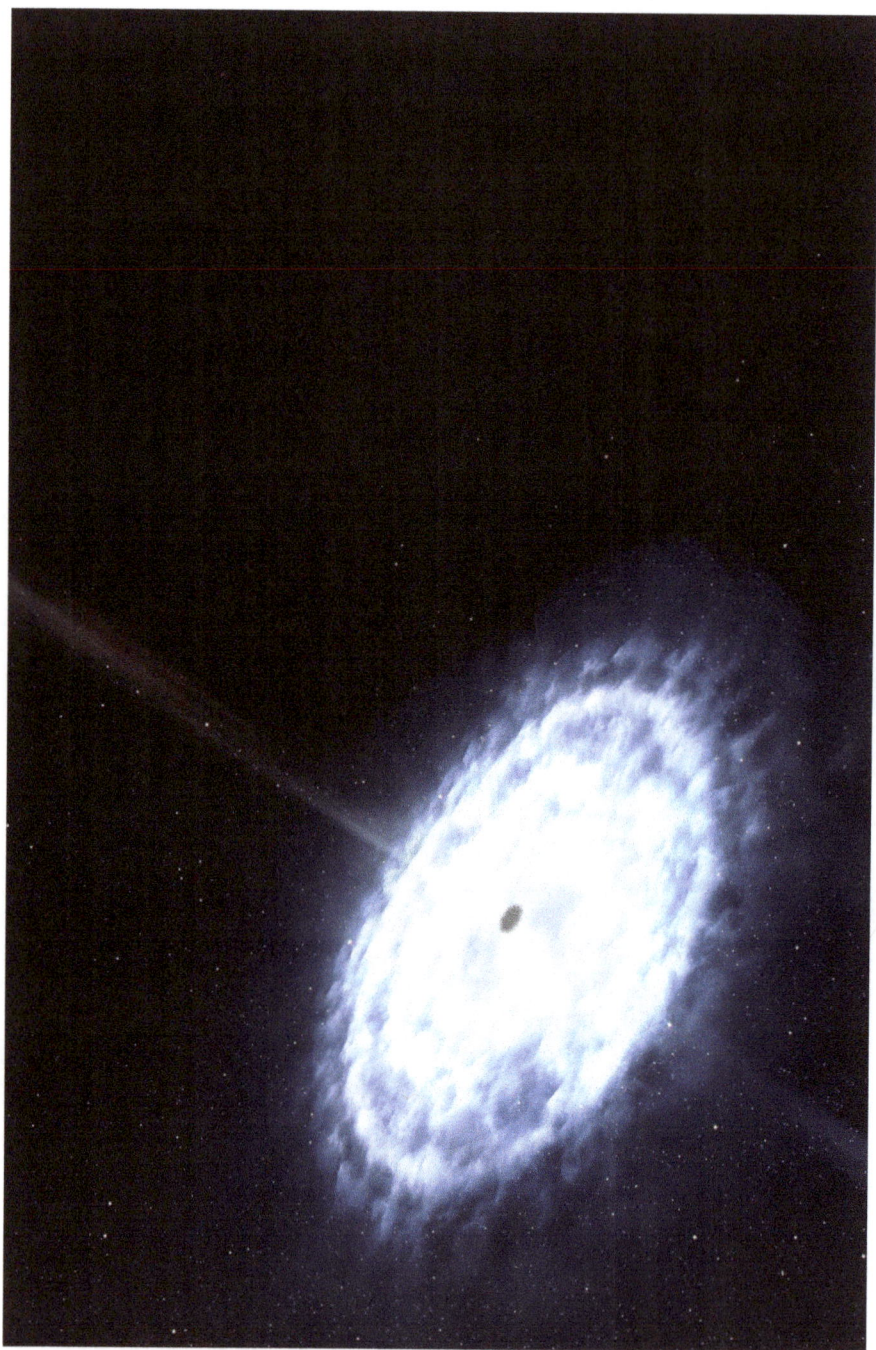

Artist impression of a black hole (only visible because of the presence of the accretion disc surrounding the tiny star).

Even if we take the position that such an experiment never happened in history, one observation that would never go missing in the U.S. military files is a scientific discussion of rendering an object invisible on an astronomical scale. In the world of astronomy, we call these invisible objects *black holes*.

Before we give an explanation of these enigmatic objects, it might be worth taking this opportunity to explain how black holes were deduced by scientists and detected in nature.

The story began in 1784 when English amateur astronomer and geologist Reverend John Michell (1724–1793) applied Newton's much-loved equations of motion to determine the escape velocity required to break free from the gravitational field of a large body. In his calculations, Michell showed that escaping Earth's gravity would require an escape velocity of one twenty-five thousandth the speed of light. To overcome the Sun's gravitational field, the figure became one five-hundredth the speed of light. Michell realized it would take a body about 500 times the mass of the Sun to give an escape velocity of exactly the speed of light.

He published his calculations in 1783 in the *Philosophical Transactions* of the Royal Society of London, volume 74, pages 35–57. The paper was titled "On the means of discovering the distance, magnitude etc. of the fixed stars".

The French scientist Pierre-Simon Laplace also independently made the same suggestion in the second volume of his 1796 edition *Exposition du Systeme du Monde* (the English translation is *The System of the World*). On page 305 of this edition he proposed that "the greatest luminous bodies of the universe are invisible". But in later editions he decided to remove this statement. Maybe his proposition was too radical to accept?

In a letter he wrote to Cambridge physicist Henry Cavendish (1731–1810), Michell concluded by saying, "all light emitted from such a body would be made to return toward it by its own proper gravity."[27]

In 1916, soon after Einstein published his General Theory of Relativity, German astronomer Dr. Karl Schwarzschild (1873–1916) noticed how Einstein's theory mathematically predicted a star's collapse

---

27   Dickinson 1988, p.82.

by its own weight if it reached a sufficient density. If the critical density is reached, the star would gravitationally collapse under its own weight, and would continue to do so until a point of infinite density is created at its center, called a *singularity*. Because of the reduction in size and the impossible increase in density, a gravitational field would exist of such great intensity would exist that no radiation, including light, could possibly escape from its luminously hot surface. Furthermore, any light approaching this object would either be absorbed by the object or bent around the object to allow an observer to see what is behind the object. The object, then, would become virtually invisible. A star equivalent to 30 Suns could, according to theoretical astronomers, end its stellar life as just such a black hole.

Schwarzschild devised a beautifully simple relationship between the mass of an object M and a radius he called the Schwarzschild radius R_s (measured in kilometers) in which the object must be compressed for the escape velocity to equal the speed of light. The mathematical relationship[28] is,

$$R_s = 3M / M_{sun}$$

where M_{sun} is the mass of the Sun.

For example, the maximum radius of a black hole formed after the collapse of a 10-solar-mass star would be:

$$R_s = 3 (10) / 1 = 30 \text{ km}$$

When Einstein learned of his theory's ability to predict singularities in space, he dismissed it as nothing more than a fanciful mathematical idea probably having no bearing on the physical and real Universe we live in.[29]

Of course, in Einstein's day, no one had detected black holes. Nowadays, scientists are confident they have found these objects, located predominantly at the center of galaxies.

So, with the fanciful idea of singularities apparently justified by observations, this has not deterred other modern-day scientists from eagerly pursuing the mathematical idea to its ultimate creative

---

28   Moore & Nicholson 1985, p.177.
29   And probably for good reason. As the density of a star increases, the amount of radiation and its frequency will increase dramatically within the star to help counteract the gravitational forces exerted on the star. A balancing act probably takes place in the gravitational forces pressing against the star and the radiation coming out, which could prevent the star from reaching infinite density at its core.

conclusion, such as raising the possibility of time travel into the past[30] and access to a multitude of other parallel and mysterious universes different from ours by traveling into a black hole. Why? It is simply because no one knows exactly what happens inside a black hole. Mathematics for describing the laws of physics break down in the so-called singularity of a black hole and start to yield all sorts of remarkable conclusions. So anything could be possible.

Nevertheless, does an object have to be compressed to infinite density to create a strong enough gravitational field to prevent light from escaping?

Not necessarily. Sure, all the preceding remarks by Michell and Schwarzschild are based on the assumption that the entire star is non-accelerating. But, of course, we all know that the individual atoms making up the star must be accelerating. How else could the star's gravitational field exist and make it compress and collapse on itself? But the assumption all along has been that the star itself does not

---

30 As physicists know, time traveling into the future is not only possible but a reality. If you travel fast enough and within the speed-of-light range, then you can return to where you started and see your older self (i.e. it will have to be your twin brother or sister, of course). However, can one reverse the direction of time and travel back into the past? It appears mathematicians have been keen to uncover all sorts of possibilities in Einstein's General Theory of Relativity since 1937. As Elizabeth Howell (http://www.space.com/21675-time-travel.html) puts it, "some scientists have extended his [Einstein's gravitational field] equations and said it might be done". The trick here is "to go faster than light". Why? Because time slows down as one approaches the speed of light. At the speed of light, time stands still. And if there is a way to exceed the speed of light, wouldn't time reverse direction and allow one to go back in time? Seems mathematically logical, but is it based on reality?

 Leaving aside the enormous technical difficulties (even for the most advanced alien civilization) of exceeding the speed of light for any ordinary object because of how it attains infinite mass at the speed of light, some mathematicians have derived solutions from Einstein's General Theory of Relativity to suggest there might be ways.

 Usually, these solutions involve setting up such extreme situations that again the question of whether it can be realistically achievable has been raised. At any rate, it has not stopped mathematicians and physicists from dreaming up mathematical possibilities just to capture the imaginations of science fiction writers and the general public.

 For example, talk of the extreme conditions around a black hole is one example. If you could reach and go slightly beyond the event horizon (the point at which the speed of light cannot escape the black hole) and somehow find a way to come back from it, you could find yourself (if the black hole rotates) back to where you started, and you may possibly see your younger self.

 But black holes are naturally hard to reach (and a good thing too, as most people would not appreciate being sucked into a black hole). So what about those theoretical wormholes predicted by Einstein's theory? Technically, a wormhole would allow anything placed inside to exceed the speed of light. And wormholes could be created anywhere in space if there was a technical way of achieving this. However, as the Unified Field Theory tells us, wormholes are nothing more than perfect vacuum regions. In other words, there is no radiation (i.e. the thing that controls space-time). Furthermore, the Universe has an infinitely powerful way of ensuring perfect vacuums do not exist anywhere. Therefore, as far as time travel into the past is concerned, this remains a theoretical idea and is probably not based on, or will ever be based on, reality. And probably for good reason too—it would break one of the fundamental rules of "cause and effect" (it is meant to be a one-way traffic flow moving forward in time). The last thing you want is someone to go back in time to affect the decisions of your parents to create you. Because if that happens, you will instantly disappear from this Universe and probably be someone else entirely.

accelerate. In the case of Michell, he believed adding extra mass (again having accelerated atoms and subatomic particles) to a star to make it bigger is all that is required to create a gravitational field sufficient to cause light to bend on itself. Schwarzschild believed that you do not need to add extra mass; you just compress the mass, and somehow, the gravitational field gets stronger. Well, to be more precise, compression simply makes the atoms accelerate more rigorously and so emit higher frequency radiation. It is the radiation that is generating a stronger gravitational field.

But did you know the entire star can spin on its axis to amplify the strength of the gravitational field? Of course you do. Spinning is just another form of acceleration.

This is an unquestionable fact of nature. Spinning is a common and natural phenomenon. Every object will undergo some form of spin. We see this everywhere. The Earth spins on its axis to give us our experience of day and night. Even the Sun and all the stars in the visible universe spin on their axis. Spin is just another way for the physicists to say that the object is accelerating (i.e., changing direction). Thus, the higher the spin, the greater the acceleration and the stronger the gravitational field will get.

Even Einstein made the spinning idea more poignant when he wrote in his famous 1905 paper on special relativity:

> "Thence we conclude that a balance clock at the equator [of the Earth] must go more slowly, by a very small amount, than a precisely similar clock situated at one of the poles under otherwise identical conditions."[31]

In Einstein's Special Theory of Relativity, he attributes this to the high speeds of the mass traveling around the equator, which increase the mass and, therefore, result in a stronger gravitational field. In Einstein's General Theory of Relativity, any ordinary mass accelerating naturally increases the curvature of space (or the gravitational field of the mass-energy of space through an increase in its energy density). In the Unified Field Theory, this increase in the density of mass-energy of space is revealed to be radiation.

---

31   Bernstein 1991, p.3.

Even if we could ignore the contribution of radiation to this gravitational field concept (because scientists think that Einstein's Unified Field Theory is too controversial to consider for the moment), we cannot ignore the contribution spin makes to the strength of a gravitational field. It is a form of acceleration, and scientists accept the General Theory of Relativity for describing how the acceleration affects the gravitational field.

Suppose for theoretical considerations that we had a spherical object with a radius of 30 km spinning at a rate of, say, 0.63 milliseconds[32] for each revolution. The question is, "How massive would this object get as perceived by an outside observer?" The answer is infinite. At this impossibly fast spin rate, the mass around the equator of the object would reach the speed of light. An outside observer looking at this impossibly rapid spin motion would measure the gravitational field around the equator at its surface as infinite. In other words, the entire object is a singularity and the field would be too powerful to imagine precisely.

In reality, we know that no object can ever produce an infinitely high gravitational field, and hence, a star cannot quite spin at the speed of light. But there is nothing that would stop an object from approaching very close to the speed of light in order to produce a powerful gravitational field. In the case of the 30-km-radius spherical object mentioned above, a spin rate of, say, 0.65 milliseconds or greater (well, anything above 0.63 milliseconds) would be perfectly fine. It just needs to spin fast enough to produce a strong enough gravitational field to help do things like hold in place a group of stars to form a galaxy.

Can stars spin in the millisecond range?

We know that in the observable universe, certain stars designated as pulsars are known to emit light pulses (presumably around the poles of the star as it spins on its axis) measured in milliseconds, so could black holes be faster-spinning stellar objects? Or are black holes merely pulsars where the spinning action does not allow the light from the poles to sweep across space to reach the Earth?

Either way, black holes could be ordinary stars spinning at a high rate to create a powerful gravitational field in order to render themselves invisible. Forget the idea of traveling to another universe in

---

32   A millisecond is one thousandth of a second.

a black hole. It is more likely closer to reality to say that if you fall into a black hole, you will be instantly ripped apart by the radiation, flattened thinner than the flattest pancake in the Universe, and burned to oblivion. Ouch.

Do the spin rates of black holes ever slow down over time?

Yes, they do. Although light and matter caught up in the gravitational field of a black hole can eventually fall into it, thereby adding extra weight and thus a stronger gravitational field, the collision of the light and matter will impinge on the spinning rate of the star by a tiny amount. As more light and mass hits the star, the kinetic energy is slowly lost or "evaporated" through the poles of the star as jets of radiating energy. As Professor Stephen Hawking, the world-famous author of *A Brief History of Time*, understood, some light or "information" can escape a black hole according to the rules of quantum physics. This leaking radiation has been named "Hawking radiation" in honor of Professor Hawking's work in revolutionizing the study of black holes.

Okay. If black holes are invisible, how do we know they exist?

The only thing that informs us of the existence of a black hole is the way its powerful gravitational field can influence matter around it. For example, at close range, matter will be swept up by the black hole's gravitational field to form a heated torus of ripped-apart material around the equator. At large distances, the gravitational field around a black hole can hold phenomenal numbers of stars, often in the hundreds of millions, around it to form a galaxy, such as our Milky Way.

Or to put it simply, scientists only need to observe visible matter sweeping around the equator of an incredibly small object to know that black holes must exist.

Also any emission of radiation from the poles of the black hole can ionize atoms in space. When the atoms regain their electrons, they emit radiation, allowing scientists to indirectly detect the existence of a black hole.

There is photographic evidence to support the existence of black holes at the centre of galaxies. For example, the accretion disc and jets of light created by a rapidly rotating star can be seen in photographs of one of the brightest galaxies in the Virgo Cluster known as NGC 4261, taken by the Hubble Space Telescope. In the photograph shown on the next page, we see a composite false color image and the original photograph image of the galaxy. The stars in this giant elliptical galaxy are shown in the standard optical spectrum (in white), whereas the massive radio jets spanning some 88,000 light years from end to end are shown in false color (in orange) to reveal the strong radio emissions emanating from the black hole.

The Hubble Telescope has also photographed the source of these radio emissions at the center of the galaxy. It reveals a dusty and superheated 400-light-year diameter torus swirling around a rapidly rotating star (as shown on this page).

Black hole at the centre of NGC 4261.

And why stop there?

It is likely that the "spinning motion" idea could be used to explain how atomic particles like electrons and protons obtain their discrete values for mass and charge.

For example, the spinning nucleus of an atom containing positively-charged protons and supposedly uncharged neutrons could be emitting radiation to keep its family of negatively-charged electrons in discrete orbits around it without falling into the nucleus. This is the familiar "synchrotron radiation". So apart from generating a recoiling force to keep the protons and neutrons inside the nucleus, which could be an important contributing factor to the strong nuclear force, stationary and concentric gravitational (or radiation) wave ripples can exist around the nucleus thanks to this synchrotron radiation of decreasing frequency the further you move away from the nucleus. These stationary gravitational wave ripples help hold the electrons in their respective orbits.

*Ground-based optical image of galaxy NGC 4261 (centre)
superimposed with radio wave image (red/orange).*

And if scientists discover that the nucleus can also render itself invisible because of the intense gravitational field generated by the synchrotron radiation, maybe we should not be surprised by this revelation. Who knows? Every atom may well hold a mini-black hole at the center.

View of Dr. Wilbert Smith

All right then. Where else can we find more evidence to support the link between the gravitational field and the electromagnetic field? Fortunately, we do find in the literature an interesting paper written by a former senior radio engineer named Dr. Wilbert B. Smith who was working at the Department of Transport in Ottawa for the Canadian government in the 1950s.

In his unpublished paper titled *Suggestions on Gravity Control through Field Manipulation*, Smith suggested that,

> "...the electric field induced by the motional magnetic field could and probably does have very much the same properties as gravity, and in fact might be the same thing."[33]

A motional magnetic field implies the presence of a motional electric field which is just another way of saying a motional or oscillating electromagnetic field. Motional just means time-varying. Whether the electric field is in fact the gravitational field is still debatable, but notice how Smith mentions the importance of a motional magnetic field. A motional magnetic field is an oscillating electromagnetic field, so a motional magnetic field must produce a gravitational field of its own if the statement from Infeld remains true (assuming what he is saying for the electromagnetic field is time-varying).

Given the evidence so far, how much more will it take for scientists to find out if the interpretation (and the picture) we now have in this book is true?

---

33   Smith, p.2.

View of Dr. Vàclav Hlavaty

Still not convinced? We could always ask Dr. Vàclav Hlavaty (1894–1969) of the University of Indiana's Graduate Institute of Advanced Mathematics, Bloomington, Indiana, USA, for his views.

One of a number of top electromagnetic experts contracted by the U.S. Government in the mid-1950s to study gravity and universal gravitation, Hlavaty has produced much enlightening work on Einstein's Unified Field Theory, including the following publications:

1. *The Elementary Basic Principles of the Unified Theory of Relativity.* Proceedings of the National Academy of Sciences, USA, Vol.38, 1952, pages 343–347.
2. *Report on the Recent Einstein Unified Field Theory.* Rendiconti del Seminario Matematico della Università di Padova, Anno XXIII, 1954, pages 316–332.

In 1957, toward the end of his contract with the U.S. Government, Hlavaty wrote the book *Geometry of Einstein's Unified Field Theory*, published in Groningen, Holland by P. Noordhoff Ltd. and supported by the National Science Foundation in the United States.

If anyone should know something about Einstein's Unified Field Theory, that person would be Hlavaty. Well, in *Geometry of Einstein's Unified Field Theory*, Hlavaty says:

"...we see that the gravitational and electromagnetic fields are interrelated."

This quote appears in the preface section of the book on page xx. Hlavaty also formally makes the relationship a theorem on page 206:

"THEOREM (21.2). The gravitational field and the electromagnetic field are interrelated. The $\left(\frac{gravitational}{electromagnetic} \right)$ field generates the $\left(\frac{gravitational}{electromagnetic} \right)$ field."[34]

Intriguing. Hlavaty is claiming beyond a shadow of a doubt that the electromagnetic field can not only affect the gravitational field, but also

---

34   Hlavaty 1957, p.206.

that one field generates the other and vice versa, such that the fields might not be independent of each other as scientists had originally thought. This goes one step further than Infeld's statement where he said, "A pure gravitational field can exist without an electromagnetic field". For Hlavaty, neither field can exist without the presence of the other.

This has to be the clearest evidence yet in support of the previous analysis that radiation probably is the gravitational field. It is all a question of whether scientists want to retain the old "gravitational field" concept or forego it altogether in favor of a purely electromagnetic Universe controlled entirely by radiation. Dr. Hlavarty has not gone quite as far as that, as he has chosen to keep the old gravitational field concept going for longer in order to remember the efforts of Sir Isaac Newton in coming up with the idea. But should we retain such outdated ideas even out of respect of a famous scientist's work?

So, does this mean Infeld was wrong? Apparently not. Hlavaty supports Infeld when he writes on page 180:

> "THEOREM (9.2). There are electromagnetic fields which do not generate gravitation. In particular, the electromagnetic field of a plane wave in the electromagnetic theory of light does not generate gravitation."[35]

A plane wave electromagnetic field is essentially a static electromagnetic field. Infeld must have been referring to static electromagnetic fields when he made his statement about the independence of the gravitational field from the electromagnetic field, In the case of Hlavaty's statement, it becomes true when we say the oscillating electromagnetic fields is the key to generating gravitational fields, and that the gravitational field must always come with an oscillating electromagnetic field.

How to Experimentally Confirm the Interpretation

With all this talk of a link between the electromagnetic field and the gravitational field, it seems befitting that we discuss how we can test this idea through an experiment.

---

35   Hlavaty 1957, p.180.

Experimental test 1

The first experimental test physicists can perform involves reducing the temperature of an object to absolute zero Kelvin—the coldest temperature known to science. At this temperature, the electromagnetic fields collapse to zero because the electric charges consisting of the electrons and protons inside the atoms comprising the object are finally brought to a complete standstill. In that case, what happens to the gravitational field? If the Unified Field Theory is correct, the field should drop to zero as well.

Readers should be aware that temperature is a measure of the quality (i.e., frequency) of the oscillating electromagnetic field in and around any object, including the fields emitted by the charged atomic particles of the atoms comprising the object. In other words, the higher the frequency of the oscillating electromagnetic field, the higher the temperature of the object itself. That is why a dull red-hot ball of metal emits radiation of lower frequencies compared to a bright super-hot orange, yellow, or bluish-white metal.

Now, what happens to the gravitational field of the object when we eliminate all the electromagnetic fields inside the object? Does the gravitational field still exist, indicating that both fields are independent and something exotic in solid ordinary matter is creating the gravitational field? Or will the gravitational field disappear with the electromagnetic field, thereby proving the two fields are not independent?

If we discover that the fields are indeed independent, reducing the temperature of the mass to the absolute lowest possible level should not affect the ability of the mass to be influenced by the gravitational field of the Earth. Such a supercooled piece of matter should effectively fall to the Earth's surface, which means it would undoubtedly prove that something else must be generating the object's own gravitational field that interacts with the gravitational field of the Earth. If, on the other hand, the fields are not independent of each other (and so proving the unified field idea), it would mean that the supercooled object should merely float in the air. No energy required. Just move the piece of supercooled matter anywhere in the atmosphere, and it will sit there thumbing its proverbial nose (supposing it has one) at the gravitational field of the Earth. Then we have what we call a true anti-gravity device.

In an act of gravitational defiance, a magnet floats above a superconducting material cooled to a specific temperature. However, if you reduce the temperature of any material to absolute zero kelvin, the Unified Field Theory predicts the material should float in any position in space (assuming the material stays intact). There is no need to have a superconductor to achieve this remarkable feat.

Experimental test 2

There is another easier way of testing the unified field idea. We could build a perfectly symmetrical Faraday cage (achievable by going into space and letting a piece of molten metal solidify into a perfect, hollow sphere). It does not matter what sort of metal is used to build the "cage", just so long as it is an excellent conductor of heat and electricity. The shape is also not too important so long as it is symmetrical. A sphere is perfectly fine and probably the easiest to build in space.

According to the laws of electromagnetism, the oscillating electromagnetic fields inside the symmetrical metal cage will cancel each other out by the out-of-phase electromagnetic fields reflected off the opposite side of the metallic wall, thereby reducing the electromagnetic fields to zero inside the cage. Scientists call this the *Faraday-cage effect*. Now, if the electromagnetic fields go down to zero, what happens to the gravitational field inside the sphere?

If a perfect metal sphere reduces the electromagnetic field inside to zero, what would happen to the gravitational field if the man is placed inside the sphere?

If the fields are not independent of each other such that one is generating the other and vice versa, then this fact should reveal itself as a weightlessness of anything placed inside the sphere. Even if the sphere is brought back to the Earth's surface, the gravitational field should remain zero because the electromagnetic field inside is zero, and so the effects of weightlessness inside the sphere should be maintained. Of course, anyone sitting inside this sphere would feel extremely cold in this environment. So you can turn on some lights inside to warm things up. But if the flow of a majority of this radiation inside the metallic sphere is not in any particular direction, then you should still experience weightlessness.

Again, this weightlessness inside the metallic sphere should take absolutely no energy to achieve. You would simply be floating in the air like you were in space.

Of course, if you don't like the idea of floating around inside a sphere, you can always create artificial gravity inside by covering the floor of the room inside the Faraday-cage box with a black radiation-absorbing material to prevent radiation from being reflected back up.

The black material simply reduces the amount and quality of the radiation going up to the roof, but radiation reflected off a silvery metal ceiling will exert a slightly greater downward force to keep things on the ground. To help increase the gravitational effect inside, consider increasing the amount of radiation coming down from the ceiling using a bright light source.

Now that isn't difficult physics after all!

Similarly, if you try to accelerate the sphere, the gravitational field should remain zero. Hence, the field cannot create *inertial forces* on anything sitting inside.

Experimental test 3

It is interesting to see how the previous two experiments are capable of proving or disproving the unified field idea with gravity's claimed link to light.

If the scientists want a little more excitement, why not try the experiment allegedly used by the U.S. Navy? Instead of reducing the electromagnetic field to zero, why not amplify the oscillating electromagnetic field? Amplifying the field (i.e., raising its energy density) should yield a powerful gravitational field capable of bending light and rendering objects placed inside this high-energy density field invisible. If invisibility does occur with an oscillating electromagnetic field, then clearly the field has to be generating a gravitational field. However, to determine with absolute certainty whether the two fields are independent or not will require an experiment involving a reduction in the electromagnetic field to zero to see how it affects the gravitational field.

It is the only way to be certain.

Should we perform the experiment?

Yes, the time has come to test the interpretation.

We should experimentally verify the interpretation of the Unified Field Theory as presented in this book because there are scientists out there who believe that gravity cannot be shielded or nullified by any means, as Suzanne Willis, acting associate dean of MadSci Network, confirms:

> "We have not observed any gravitational shielding or negative gravitational effects in nature."[36]

However, it may end up being closer to the truth to say that the oscillating electromagnetic field is, in fact, controlling gravity. In other words, if you do anything to nullify the electromagnetic field through a process known to the scientists called *destructive interference* (by combining two out-of-phase electromagnetic waves), you should be able to affect the gravitational field (i.e., nullify it too). Likewise, raise the density of the electromagnetic field by increasing its frequency and/or electric charge, and the gravitational field should go up as well. That is the link we must focus on.

Surely this idea must be something worth exploring…and testing.

The Speed of Light is the Speed of Gravity

Still can't believe the link is true? We understand how significant and radical this paradigm shift is for our readers. It requires us to realize light is probably gravity, and if we push the boundaries on this concept, we may not even have a gravitational field. There is every possibility we could live in a purely electromagnetic Universe. So to help further hit home the concept of the Unified Field Theory, let us present two more pieces of evidence.

This first piece of evidence concerns the speed of a gravitational wave. Started by Einstein and later reinforced by other scientists, the public has been informed that gravitational waves propagate at the speed of light. Well, hold on! Has anyone noticed something really odd here? If we think about it, this happens to be the exact same speed of the oscillating electromagnetic field. Huh? Light and gravity traveling at the same speed? What a coincidence! Why would these fields have to travel at the *exact* same speed if they are meant to be separate forces of nature? Now, for the first time, an experiment has started the ball rolling by confirming the speed of gravity.

In an article published in the *New Scientist* magazine dated January 7, 2003, and titled "First Speed of Gravity Measurement Revealed", two

---

36   Quote obtained from http://www.madsci.org/posts/archives/2005-04/1112889896.Ph.r.html.

scientists Ed Fomalont of the National Radio Astronomy Observatory in Charlottesville, Virginia, and Sergei Kopeikin of the University of Missouri in Columbia, made the first measurement on the speed of gravity with the help of the planet Jupiter.

A true color picture of Jupiter as seen by NASA's Cassini spacecraft, taken on December 29, 2000.

The experiment involved scientists looking at the radio waves from a distant quasar focused into a ring and seeing it get distorted as the radio waves interact with Jupiter's own gravitational waves.

First, Kopeikin managed to rework the equations of general relativity in such a way as to show that the gravitational field of a moving body can be determined by just three basic variables: mass, velocity, and the speed of gravity. Since scientists already know the mass and velocity of, say, a large planet (in this case, Jupiter), it should be a relatively simple matter to work out its gravitational field and, with it, the speed of gravity. Second, the scientists had to choose a distant radio wave source (in this case a quasar) to conduct the experiment.

In September 2002, Kopeikin and his colleague Fomalont had the opportunity to test the speed of gravity by observing the radio waves from an extremely distant and bright quasar as they bent slightly around Jupiter and became distorted by its gravitational waves of the planet before the radio waves finally reached the Earth.

After experimentally measuring the apparent change in position of the quasar, the scientists were able to work out the gravitational field of Jupiter and, with it, provide the world's first measurement of the speed of gravity.

The official figure showed a speed of 0.95 times that of light with an error margin of plus or minus 0.25—well within the expected range needed to support Einstein's assumption that the speed of gravity is the speed of light. As the scientists were quoted in the *New Scientist* magazine:

> "We became the first two people to know the speed of gravity, one of the fundamental constants of nature."

The result was announced at a meeting of the American Astronomical Society in Seattle, USA.

Light Has Same Range of Influence as Gravity

The second piece of uncanny evidence is how the electromagnetic field just happens to have the same range of influence over interplanetary, interstellar, and intergalactic distances as the gravitational field. We can see this from the table below (the data is available in any good university physics textbook):

FORCE	RANGE	RELATIVE STRENGTH	ROLE IN THE UNIVERSE
Strong	10^{-15} meters	1	Holds together the quarks inside protons and neutrons
Electromagnetic	Infinity	1/137	Determines structure of atoms, molecules, solids, and liquids.
Weak	10^{-17} meters	10^{-6}	Determines stability of atomic nuclei; fuels the Sun and stars.
Gravitational	Infinity	6×10^{-39}	Assembles large quantities of matter into planets, stars, and galaxies.

As the table shows, of all the known forces of nature, the strong and the weak forces are the least likely to control gravity because they affect only atomic and subatomic particles over a very short distance (although the forces could still be electromagnetic in character or purely gravitational if you want to keep the latter field alive; it all depends on how powerful you like to make the electromagnetic or gravitational fields to duplicate the strength shown by these more exotic nuclear forces[37]). Unless there is a more exotic particle or force of nature not known to science that is responsible for controlling the gravitational field over much greater distances, we are pretty much forced to conclude that the only thing that has a chance of controlling the gravitational field over the same distance is the electromagnetic field.

Given what we now know, what is the probability that the gravitational field isn't the oscillating electromagnetic field (i.e., gravity isn't light)? As they say, "If it quacks, has feathers, and comes with a beak and webbed feet, shouldn't we call it a duck?"

Actually, it could even be the Donald Duck we mentioned earlier in this chapter.

Or should we consider it yet another coincidence in the bag of coincidences?

---

37   Simply by raising the frequency (or energy density) of the electromagnetic field.

Perhaps it is time we ignore our historical sentiment in the gravitational field as an enduring legacy of the great work of Sir Isaac Newton so many centuries ago. Let bygones be bygones. It is time we all move on and start afresh. And this time, let us think in a purely electromagnetic way.

Can We Explain Other Mysteries of Science?

Still unsure? As with all controversial ideas, it is okay to feel a little skeptical. That's the way science works. So let us see if we can find more evidence to support the link.

For example, suppose there is a way to explain many of the mysteries of the visible universe in a purely electromagnetic and classical Newtonian way with light playing the central role. Would these new electromagnetic explanations give greater credence to Einstein's Unified Field concept?

To see if this is true, in the remaining chapters of this book, we will present a new electromagnetic interpretation for how universal gravitation and gravity may work. In quantum mechanics, we will use light's "particle-like" or "gravitational-like" property to explain the wave-particle duality of quantum particles as well as other seemingly unique quantum phenomena such as *quantum entanglement*. In astronomy, we will tackle the question of the age and size of the Universe. We will also figure out whether the aging process for all biological systems could have its basis in electromagnetism in terms of the impact of radiation on living cells. And, most controversially, we will determine in the final chapter whether light can provide insights into what God might be.

CHAPTER 4

What Is Gravity?

Imagination is more important than knowledge. For knowledge is limited to all we now know and understand, while imagination embraces the entire world, and all there ever will be to know and understand.

—Albert Einstein[1]

EVER SINCE Sir Isaac Newton discovered gravity and universal gravitation, it has always been interpreted as an apparent mutual force of attraction *pulling* matter together. But what if gravity is more of a *pushing* action?

---

1   This is a variant on the original quote from Einstein published in his 1931 book, *Cosmic Religion: With Other Opinions and Aphorisms*, on page 97: "I believe in intuition and inspiration. ... At times I feel certain I am right while not knowing the reason. When the eclipse of 1919 confirmed my intuition, I was not in the least surprised. In fact I would have been astonished had it turned out otherwise. Imagination is more important than knowledge. For knowledge is limited, whereas imagination embraces the entire world, stimulating progress, giving birth to evolution. It is, strictly speaking, a real factor in scientific research."
   He also made another variant on this quote during an interview that was published in *The Saturday Evening Post* in 1929 when he said: "I am enough of the artist to draw freely upon my imagination. Imagination is more important than knowledge. Knowledge is limited. Imagination encircles the world."

Portrait of Sir Isaac Newton in 1689.

Why Does an Uncharged Object Float in Space?

Author and researcher John A. Keel has stated the following indisputable fact about this Universe[2]:

> "Electromagnetic waves of many different frequencies permeate the known universe. We live in a sea of such radiations, and the space through which our planet travels is an ocean of radiation."[3]

---

2   See Chapter 6 for why the Universe is capitalized.
3   Keel 1970, p.57

As we are literally swimming in a sea of radiation, and knowing that the Unified Field Theory requires us to look at the Universe from an electromagnetic perspective, it is only reasonable that we continue using radiation to see how we can create a new theory of gravity and universal gravitation. In particular, we need to examine how solid matter affects the radiation around it and, in turn, determine how this might affect the force of the radiation on solid matter. In that way, we can determine how likely matter clumps together in the presence of radiation.

Let us imagine a spherical, uncharged object immersed in this sea of radiation. The first thing we would notice is how radiation constantly bombards the uncharged object from all directions[4]. As the amount and quality (or frequency) of the electromagnetic energy from space and is being absorbed and emitted by the electrons in the atoms comprising the object are the same on average all around the object, the energy will naturally push the object, causing it to recoil in the opposite direction. However, because this happens all around the object with equal force, and because the atoms are held in place by the chemical bonds forming the object's crystalline structure, the entire object will not appear to move. It will simply float in space. Of course, using a power microscope will undoubtedly show that the atoms and the object itself naturally vibrate (i.e., constantly accelerate) ever so slightly in all directions. But let us not worry about this kind of fine detail for the moment.

Increasing the temperature of the object will not cause the object to move, either. All the temperature does is agitate and cause greater vibrational motions in the atoms leading to emission of more radiation of a higher frequency from the atoms and the object's surface. So,

---

4 Scientists may loosely describe this radiation as based on "temperature". True, temperature does require radiation to be present. However, temperature in this context has a narrow meaning. Temperature usually refers to radiation in the infrared range leading to the sensation of heat (i.e., how hot or cold something is). Scientists may include microwave radiation in this definition because of its ability to agitate molecules, which in turn leads to the emission of mostly infrared energy. At other frequencies outside the microwave and infrared regions, the amount of infrared energy emitted by molecules when collisions take place with the radiation will vary significantly. In this electromagnetic theory of gravity, we define "radiation" to include all types, from the lowest to the highest frequencies. Any mathematical modeling performed by scientists to test this theory must take into account all forms of radiation and must also include what happens to radiation (in terms of its speed and, ultimately, the pressure it exerts on ordinary matter) when the mass-energy density of this radiation in space is reduced by any suitable means. Scientists should also take great care not to assume that the mass of a single photon is so small that they can treat it as negligible and, therefore, approximate it as such in the mathematical modeling. The totality of all such mass in the radiation permeating the Universe should be carefully considered.

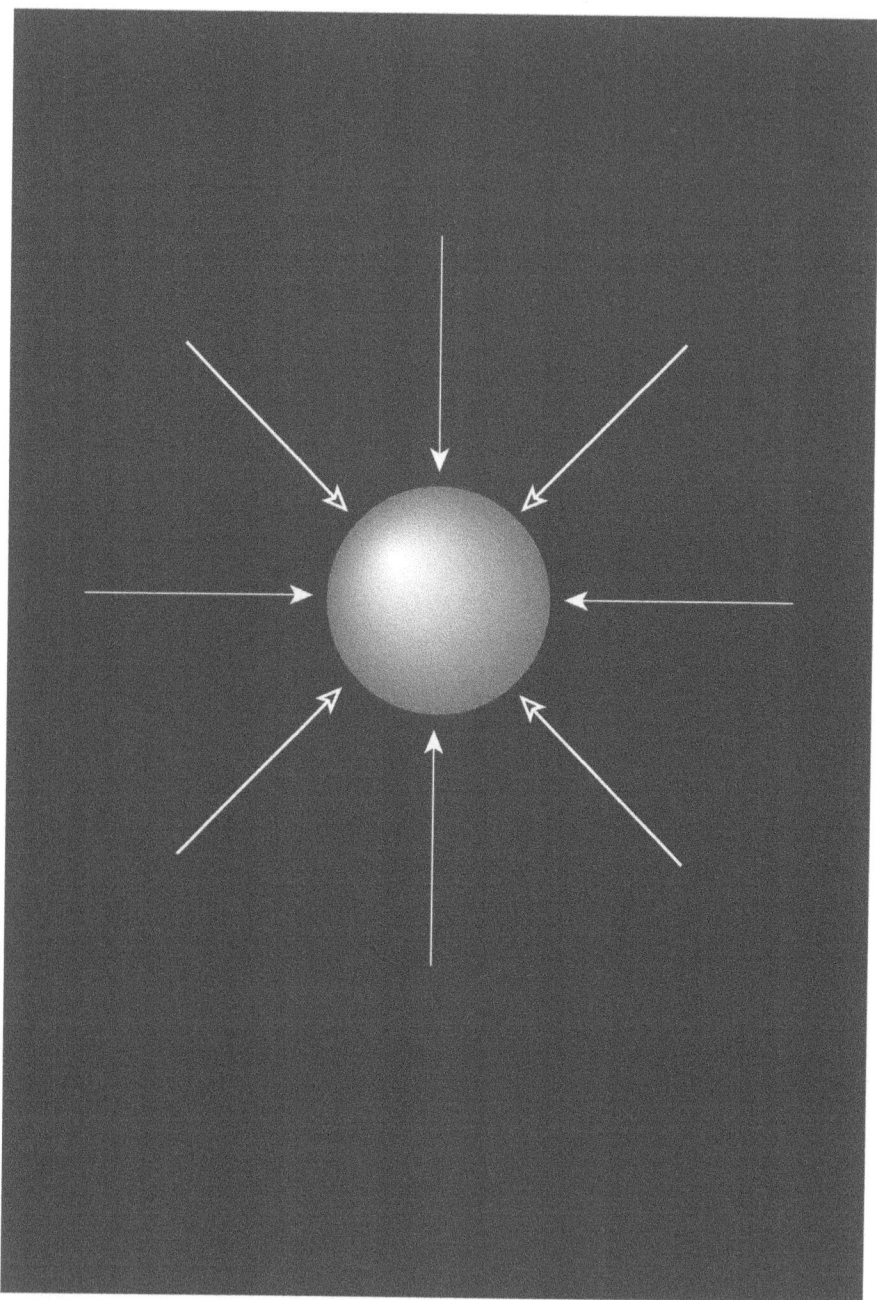

According to the Unified Field Theory, uncharged objects float in space because the universal background radiation is exerting an equal force all around the object.

because the emission is roughly equal all around the object, it will continue to float in space.

Likewise, if you spin the object to create greater acceleration, more radiation gets absorbed and emitted. But again the energy density of the radiation is the same all around the object's equator. There might be a different energy density at the poles, but nothing to cause the object to move. With no perceptible variation in the quality (or frequency) and quantity of radiation at opposite ends of the object, there is no propensity for the object to move in any one direction. As such, the object will appear to float in space in a process known as "weightlessness".

Similarly, if radiation in space were to disappear, the object would also not move. The only difference in this situation is that there would be absolutely no vibrations or other accelerating motions to be seen anywhere, even at the atomic level. Furthermore, the object would be completely and absolutely at rest with respect to any point in the Universe.[5]

In a perfect vacuum of space where no radiation exists, an object has no gravitational field. Even if another object is added to the Universe at any distance from the first, nothing will happen to the objects. They will remain floating as if they are independent frames of reference.

Presence of a Second Body of Mass

Suddenly, in our radiation-filled Universe, we introduce another uncharged object into the Universe. What do you think will happen next?

If the radiation did not exist, and if the objects did not emit radiation to create their own so-called "gravitational fields", then the objects would remain absolutely still. You can bring the two objects as close as you like, but nothing will ever happen. However, if you switch on the radiation in the Universe, then something interesting does happen. The objects will start to move together.

Why?

---

5   Actually, it is questionable whether matter can exist in a perfect vacuum. There is the likelihood that the perfect vacuum will tear apart the atoms and subatomic particles to release their internal energy to ensure that the perfect vacuum region is filled with energy. But let us assume the object can remain intact.

In the traditional gravitational Universe, physicists would have to explain the movement in terms of the way the mysterious force from the gravitational fields of both objects are interacting such that the gravitational forces exerted by each object on the other helps to "pull" the objects together.

Yet, according to the Unified Field Theory, we must see the Universe in a purely electromagnetic way—where radiation is now the prime mover. In other words, radiation is acting as ordinary matter to "push" each object together. Not only that, but something has changed the mass-energy density of the radiation around each object to produce movement. The only way this is possible is if the energy density of the radiation in the region between the two objects has gone down. But exactly how?

The clue lies in the way each object acts as a radiation shield for the other.

We discover from our electromagnetic Universe how the inner surfaces of the two objects facing each other receive slightly less radiation from the Universe than the sides facing away into space. This is due to the radiation-shielding effect of the mass of each object and the effect of each on the amount of radiation received by the other object. Generally, the more mass there is, the greater the shielding effect, which in turn reduces the amount and quality of the radiation available between the two objects to push them apart. Since the energy density of the radiation is slightly less between the two objects, and because radiation is ordinary matter in the sense that it has the ability to move matter, the excess or higher energy density of the radiation on the sides facing into space will have to *push* the two objects closer together. In other words, there must be movement of the two bodies toward each other in the presence of radiation.

Furthermore, as the objects move closer together, the greater is the area of shielding. Hence, the shielding effect is more effective. By "effective" we mean that less radiation is able to fill the space between the two objects. Therefore, the "pushing" force will appear to get stronger (even though it remains the same on the outside). As a consequence of this greater difference in the energy density on the inner and outer surfaces of the two objects, the objects will accelerate toward each other.

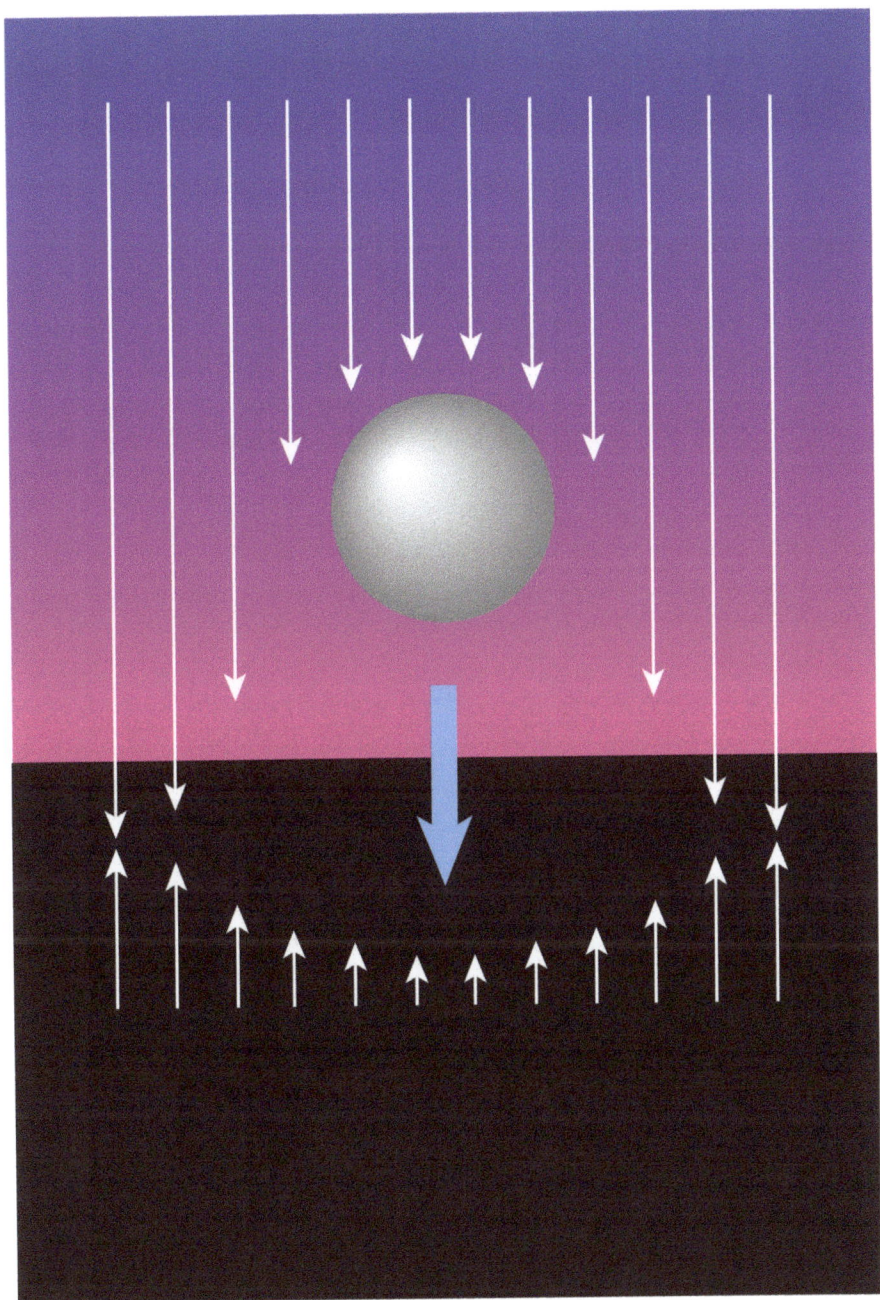

Gravity is explained by the Unified Field Theory as an imbalance in the pushing force of radiation due to radiation shielding by both objects on each other.

Well, here is the picture we are looking for. This, according to the Unified Field Theory, is the electromagnetic theory for gravity and universal gravitation.

Effect of Charge on the "Gravitational Effect"

Okay. So, what happens when the objects are electrically charged?

In the case of two charged objects, this so-called "gravitational force" caused by radiation pushing the two objects together can be amplified or de-amplified electromagnetically. It means that either the force of "attraction" can appear to accelerate more significantly or the objects can "repel" each other as if the two objects are experiencing "anti-gravity".

How does this work?

It again has everything to do with the energy density of the radiation around the objects and how we can control it through constructive and destructive interference of the radiation.

Let us explain.

Suppose you electrically charge the objects. At the very least, you will increase the amount of radiation emitted by the objects due to the presence of extra vibrating electrons (for a negative charge) or by exposing more of the vibrating protons in the atoms' nuclei (for a positive charge). At the same time, the energy emitted by each object can constructively or destructively interfere with each other, especially in between the two objects. This means the energy density in the region between the two objects facing each other can either be made to increase or decrease, respectively.

For readers unfamiliar with what we mean by the "energy being constructively and destructively interfered with each other", imagine a pond with no ripples or waves. If you get two people at opposite ends of the pond to create a single splash in the water, you will see the surface of the water move up and down forming waves (denser regions of extra water being raised above the normal surface level of the pond) and troughs (less dense and reduced quantities of water regions whose surface lies below the normal pond surface level). These waves and troughs move away from the splash sources. If you look carefully at the center of the pond, where the waves and troughs from opposite ends of the pond meet, you will notice the way two waves combine and

True anti-gravity according to the Unified Field Theory is when the energy density of the radiation between two objects exceeds the natural density of the background radiation, causing the objects to be pushed apart by the radiation.

double in height. We call this the "constructive interference of the wave". In the case of troughs behind the waves, any other wave combining with this trough will reduce the energy and the water displacement above the pond's surface in the wave is diminished and brought back to the pond's normal surface level (or lower). We call this the "destructive interference of the waves".

The same is true for radiation.

Thus, the way to control this energy density (in terms of increasing or decreasing it) in the radiation is to apply constructive or destructive interference of the electromagnetic waves. Control merely requires electrically charging the objects and bringing the objects within reasonable range.

For example, should the objects have like charges (that is, two positives or two negatives), the radiation between them will be constructively interfered with to create large and invisible matter comprised entirely of pure electromagnetic energy of a higher energy density than the rest of space. This is effectively an invisible third object (i.e., a photon) positioned between the two objects. Its presence is what stops the two charged objects from coming together. If you try to bring the objects closer together, they will be "repelled" by this third object created by radiation. It will act like a spring pushing the two objects apart.

Conversely, if you give the two objects unlike charges (that is, a positive and a negative), the radiation between them is destructively interfered, resulting in a radiation vacuum (or very close to it) between the two objects. With less radiation between the objects, the objects come together more rapidly. Why? It is because the radiation being emitted on the outside of the objects facing into space creates a recoiling force, and the radiation coming in from space collides with the objects. When we combine these forces created by the radiation, they simply *push* the objects together. With nothing in the middle of the objects to counteract and balance this situation, it means the objects will appear to be attracted to each other more significantly by the opposite charges. In reality, the radiation is pushing the objects together in a more significant way.

Additionally, as the objects are brought closer together, a greater radiation vacuum forms between them, which means the radiation on the outside pushes the objects together more effectively with nothing to counteract this force in the middle of the two objects. Thus, the charged objects will appear to accelerate toward each other the closer they are together.

In essence, the electrostatic charging of the two objects is merely performing an amplification or de-amplification of the so-called "gravitational effect".

Evidence to Support Gravity as a Pushing Force

View of Eugene Gluhareff

Perhaps the clearest support for such a radical new theory for gravity and universal gravitation can be seen in the quote from Eugene M. Gluhareff.

Gluhareff, founder and former president of Gluhareff Helicopter and Airplane Corporation of Manhattan Beach, California, USA, has given support for this new theory of gravity and universal gravitation. In an article published in *Product Engineering* titled "Electrogravitics: Science or Daydream?" Gluhareff stated:

> "E. M. Gluhareff, Pres. of Gluhareff Helicopters, suggests much progress might come if gravity were considered as "push" rather than "pull"—with all matter being pushed toward the centre of the earth by a sort of "electronic rain" from outer space."[6]

If "electronic rain" is radiation, then this is direct support for the new picture of how gravity and universal gravitation works from an electromagnetic perspective.

The Casimir effect

To further support this new electromagnetic picture of gravity and universal gravitation, there is an interesting observation concerning two metal plates brought close together (but not actually touching). The

---

6   "Electrogravitics: Science or daydream?" (December 30, 1957). *Product Engineering*, Volume 28 Number 26, p.12.

capacitor formed by these two plates helps to create what is known as the *Casimir effect*.

The Casimir effect was first predicted in 1948 by Dutch physicist Hendrick Casimir (1909–2000). The effect is nothing more than the plates moving towards each other due to the natural radiation outside the plates pushing them together. At the same time, the aforementioned shielding effect we were discussing earlier with ordinary "uncharged" matter is actually increased because the plates themselves create a Faraday-cage effect in which the emitted radiation between them is reflected back on itself in an out-of-phase manner, and this effectively causes destructive interference of the radiation. This results in a significant reduction in the energy density of the radiation between the plates. When this occurs, the excess radiation (i.e., of a higher energy density) outside the plates *pushes* them together.

This interesting effect would not be apparent (or would be reduced significantly) if the plates were not made of metal, or the entire environment containing these plates have reduced radiation to lower the density of mass-energy of space. But because the plates are made of metal (but are not electrically charged), the radiation between the plates is reduced significantly, allowing the normal radiation pressure in the environment to push the two plates together in a highly noticeable way.

Do We Really Have a Gravitational Field?

As Dr. Hlavaty, the electromagnetic expert in the 1950s, said:

> "The fact that electromagnetic field has to be present in the unified theory...makes of it a kind of *primary field existing a priori* in the frame of the unified field theory. Then we may say...*that gravitation is generated by electromagnetism.*"[7]

Of course, the problem is how to get electromagnetism to generate gravitation. Now at last, we can see what Hlavaty meant. Light could actually be generating the force of gravitation because it is universal gravitation and gravity. To put it another way, there is no distinction between light and the way ordinary matter appears to "clump together" in a gravitational way.

---

7   Hlavaty 1957, pp.179-180 (italics as in the original).

Dr Vaclav Hlavaty, professor of mathematics at Indiana University during the period 1948 to 1962.

What does this mean for the gravitational field? In particular, why do we need a gravitational field if the electromagnetic field can potentially do all the work?

We are facing the possibility that our long-held traditional Newtonian concept of a gravitational field around objects could be a figment of our imaginations, an idea having no bearing in the real Universe. The gravitational field may be nothing more than a fanciful human concept, a kind of theoretical scaffolding erected by Newton to begin the process of understanding why objects appear to be attracted to each other. Now, in the 21st century, this so-called mysterious force of "attraction" may be nothing more than radiation *pushing* matter together.

Maybe it is time to remove our gravitational scaffolding and admire the true electromagnetic masterpiece that is holding the Universe together.

CHAPTER 5

Why Do Quantum Particles Show Wave-like Behaviors

We can't solve problems by using the same kind of thinking we used when we created them.

—Albert Einstein[1]

ONE OF the most important and intriguing scientific discoveries to be made in the 20th century concerns the mystery of the wave-like behavior of matter, especially at the diminutive scale.

---

1   Quote is thought to be attributed to Einstein but when did he made this statement is unknown. An earlier variant of this comes from *The Journal of Transpersonal Psychology, Volumes 1-4*, 1969, p.124, claiming it came from Einstein: "The world that we have made as a result of the level of thinking we have done thus far creates problems that we cannot solve at the same level as the level we created them at."

Thomas Young

For nearly 80 years, the long sought-after classical explanation for this strange behavior remained elusive to physicists. But now, thanks to Einstein's Unified Field Theory, an explanation could be within our grasp.

The Thomas Young Experiment

In 1801, an avid British scientist named Thomas Young (1773–1829) conducted an experiment on the wave-like property of electromagnetic radiation, known specifically as "wave interference". Young used a monochromatic light source to send electromagnetic waves at a partition with one or more narrow slits and counted the number of bright bands he could see on a screen positioned not far behind the partition.

The first experiment Young performed was to use a single narrow slit. As he allowed the light to pass through the slit, he observed a single bright band on the screen.

"Nothing unusual here," Young must have thought.

He then used the same monochromatic light source to produce two individual electromagnetic wave sources by means of two narrow, parallel slits on a partition. The result of doing this is an overlapping of the light waves in a constructive and destructive manner forming bright and dark bands, respectively, on the screen. Young interpreted this observation, as do other scientists today, as evidence of light exhibiting a wave-like behavior. A fair-enough explanation considering the same result can be duplicated by throwing two stones in a pond and watching the ripples interact to produce larger and smaller waves through constructive and destructive interference with the waves, respectively.

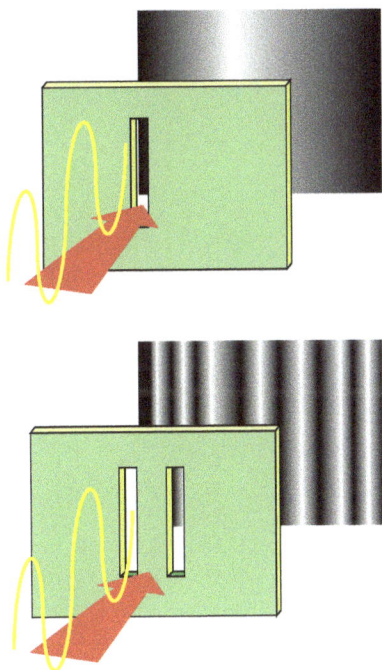

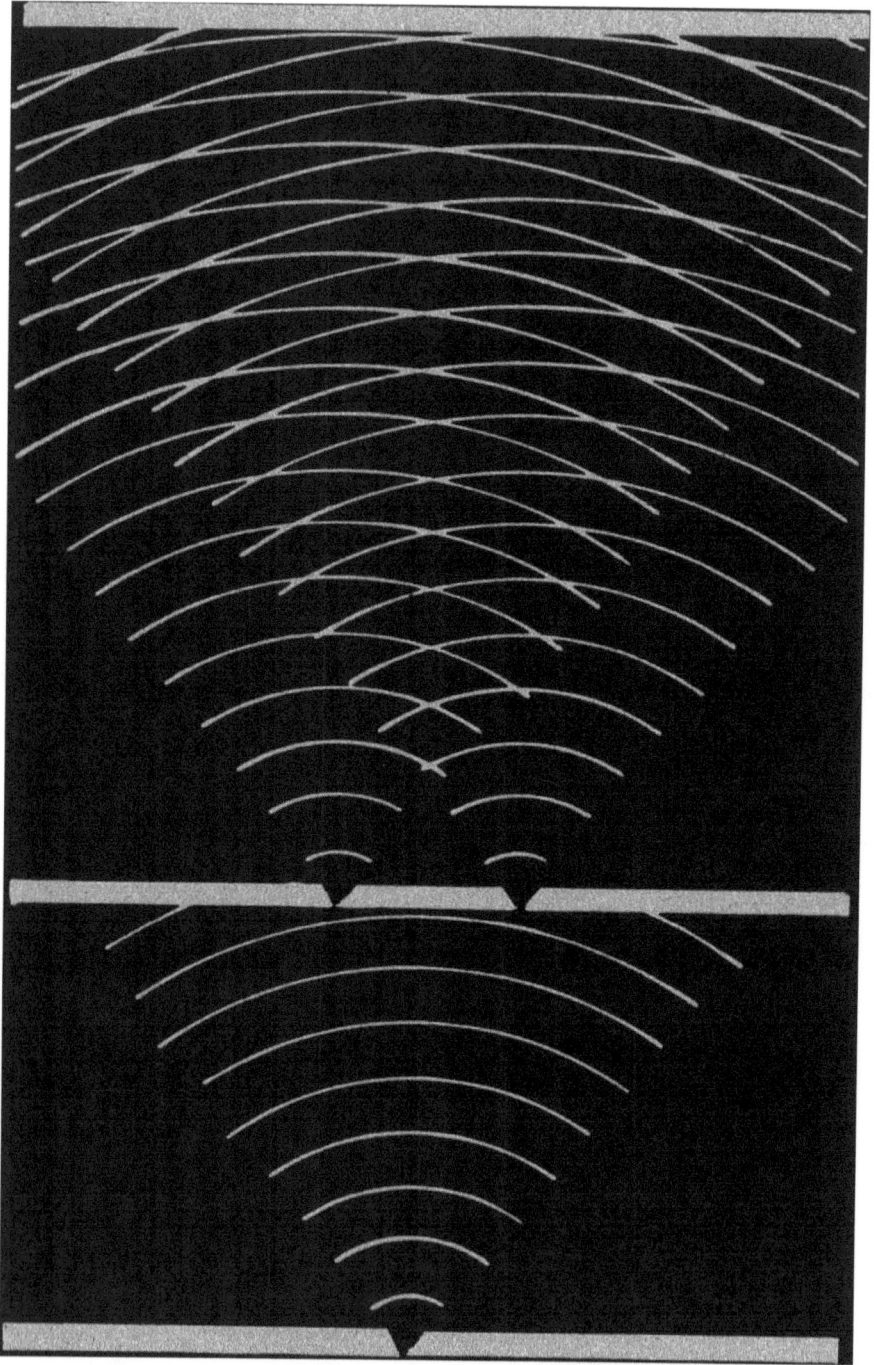

Thomas Young experiment showing light waves passing through two slits in a wall to create wave-like bands on a screen.

However, life was about to get a little more complicated for the scientists.

In the early 1920s, some creative blighter decided to repeat the experiment using electrons instead of light waves, and the entire apparatus was placed inside a vacuum chamber to remove the air molecules (but not the radiation). The modified Thomas Young experiment showed that when a single slit was present in the partition, the electrons would bunch up on the screen directly behind the slit, proving beyond reasonable doubt that the electrons were behaving like particles. This is exactly as the physicists had expected. However, when two slits were present, the result was, incredibly enough, a wave interference pattern. Even if protons, neutrons, or hydrogen atoms were used instead of electrons, the wave-like pattern would still be duplicated on the screen.

How astonishing!

Interestingly the interference pattern does not occur if the particles are very large and heavy, behaving as they should like real particles. When the particles are sufficiently lightweight, however, as soon as at least two slits are open, the particles are somehow able to change to a wave-like behavior. Quite an astonishing feat.

How can this be possible? Does this mean the particles transformed themselves into waves as soon as two or more slits were opened? One would hope not. Such a transformation would have caused the entire experiment to be destroyed in one massive nuclear explosion as the particles turned into electromagnetic waves, and there would be no one around to explain what happened. Of course, we know this is not true. The particles remained as particles. In fact, careful analysis of the bright bands or fringes revealed that the pattern was built from a myriad of tiny dots caused by the arrival of the particles (see the five images on the previous page). So how did these lightweight particles "know" how to arrange themselves in such an orderly, wave-like fashion? It is almost

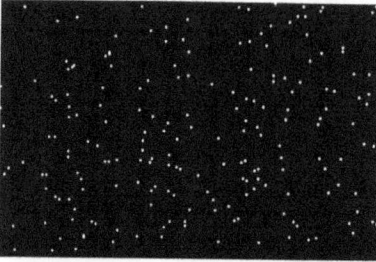

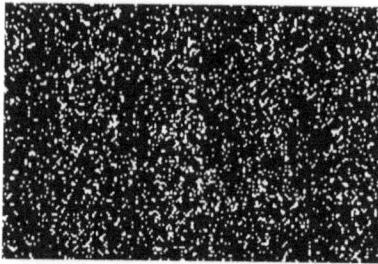

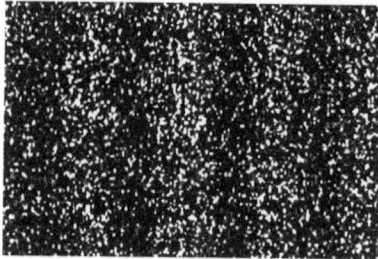

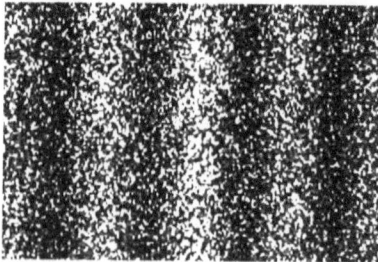

as if these particles have minds of their own. Or is this meant to be irrefutable evidence for the existence of God influencing these particles to behave in this way? Somehow one gets the feeling scientists will not agree. Yet this seemingly simple discovery has apparently astounded physicists.

As Professor Stephen W. Hawking explained:

> "The remarkable thing is that one gets exactly the same kind of fringes [in the double-slit experiment] if one replaces the source of light by a source of particles such as electrons with a definite speed (this means that the corresponding waves have a definite length)."[2]

How is this possible?

Scientists explain this phenomenon by making the claim that the particles managed to pass through both slits at the same time, according to the mathematics of quantum mechanics applied to this problem. As Hawking said:

> "If electrons are sent through the slits one at a time, one would expect

each to pass through one slit or the other, and so behave just as if the slit it passed through were the only one there—giving a uniform distribution on the screen. In reality, however, even when the electrons are sent one at a time, the fringes still appear. Each electron, therefore, must be passing through *both* slits at the same time!"[3]

However, as we know, any attempt by the particles to split apart when passing through the slits would only create a massive nuclear explosion. If that is not enough, the laws of thermodynamics are unlikely to favor the energy coming together again after passing through the slits to create the particle before landing on the screen. Too much energy would be required to put the parts together again. As there is no such explosion detected, the reality tells us that it does not matter which slit the particle enters. It must go through one slit or the other, but never both at the same time. It seems that once it passes the slits, it enters a region of space that tells the particle where to go in order to land in an orderly fashion. Without knowing what is in the space between the slits and the screen to achieve this miraculous control of the particle, we have to assume these particles have minds of their own—or God does exist and is influencing the experiment at the right times.

This is the fundamental problem of quantum theory. How do we explain this phenomenon?

We have to remember that the idea of a particle passing through both slits simultaneously is, in fact, the literal interpretation given to the mathematical solution of Schrödinger's equation (the formula at the heart of quantum mechanics) set up for this modified Young's experiment. However, mathematics can be a little too simplistic. Either the equations are set up with limited variables, or scientists oversimplify the calculations to find a solution. In the case of limited variables, it will be necessary to do a bit more thinking to see precisely what is going on and to get all the factors considered and included in the equations before any meaningful solutions can be obtained. In other words, we need to get the picture right before we can apply mathematics to solve the problem.

---

3   Hawking 1988, p.63.

This raises the prospect that there could be a classical physics explanation for the Thomas Young experiment and how particles behave as waves.

Any chance this explanation could be entrenched in the laws of electromagnetism?

Despite such promise, as U.S. physicist and Nobel laureate Dr. Richard Feynman once said:

"...[the modified Young experiment is] a phenomenon which is impossible, absolutely impossible, to explain in any classical way, and which has in it the heart of quantum mechanics."[4]

Is this really true?

Impact of Radiation on the Modified Thomas Young Experiment

Dr Richard Feynman

The first thing we need to realize is that in the original, unmodified Thomas Young experiment, the wave-like property of electromagnetic radiation was detectable only because the two individual electromagnetic waves passing through the slits had a frequency that was *visible*, either to the naked eye or to the photographic plate positioned behind the slits when showing the wave-like pattern. If our eyes and the photographic plate were not sensitive to this frequency of the electromagnetic waves, would anyone know the electromagnetic waves were there or had a wave-like property? The alternating bright and dark bands may not be visible, but that does not mean they do not exist. Now throw a bunch of electrons at the partition containing the double-slits, and it might seem astounding for us to observe wave-like behavior in these particles.

Astounding? Not really.

4   Polkinghorne 1990, p.34.

Suppose we ask ourselves: What exactly could the invisible radiation be doing to the particles in the space between the slits and the screen to make them behave in a wave-like manner?

In other words, how do we know the radiation is not influencing the particles?

Furthermore, the physicists involved claimed they conducted the experiment inside a vacuum chamber to remove the air. Very good. Glad to see someone is thinking laterally on this. But there is just one tiny problem—it was not inside a Faraday cage to annihilate the radiation, was it? As a result of this minor oversight, we still have radiation passing through the double-slits, do we not? Although the radiation is still present and passing through the slits of a certain frequency set by the width of the slits, there is the possibility that this energy is influencing the particles and guiding them to their final destinations on the screen, which just happens to result in the familiar wave-like bands.

It is like a freeway of bumper-to-bumper cars moving at high speeds along a stationary highway. You have a car, and you want to get across the highway to reach the other side. Looking at the situation in front of you, it is clear the cars on the highway are too close together and will not give you the space you need. Yet somehow you have to go through the stream of traffic. Depending on your car's speed and initial mass, however, it may be able to smash its way through those cars to get you to the other side of the highway. This might be a winning strategy if your car is large and heavy and/or traveling at high speed. But if your car is too lightweight and not moving fast enough, the traffic moving along the highway will smash into it and force it to move in the same direction as the general flow of traffic.

In the case of the double-slit experiment, the cars traveling on highway are the radiation, and the car about to enter the highway is an electron. As the electron enters a region containing extra radiation (a place of higher energy density created by the constructive interference of two electromagnetic waves), the radiation could easily push the electron in a different direction, corresponding more closely to the direction in which the radiation is generally moving. Furthermore, if the radiation is already behaving in a wave-like manner and creating

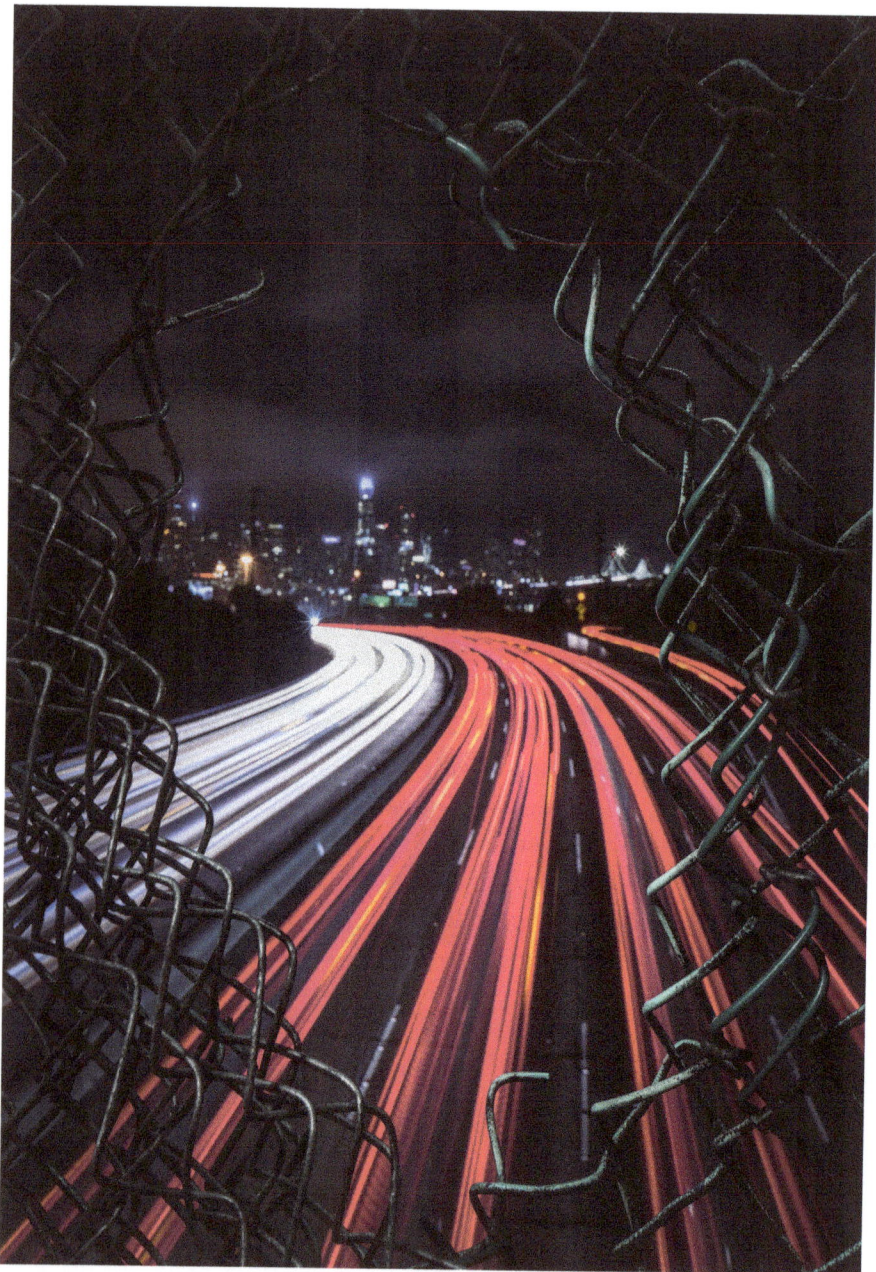

Moving cars on a road are a good analogy of photons moving in stationary light ripples. These can guide other particles to move along certain directions if they enter these regions.

bands on the screen to support this, the electron could quite easily follow this wave-like pattern.

If we want to retain the gravitational field concept to help explain why this wave-like behavior happens to solid particles such as electrons, then what we have here is intensified radiation creating stationary invisible "gravitational" wave ripples (i.e., the highway) in space. As soon as a quantum particle of sufficiently light weight and traveling at a sufficiently slow speed enters one slit or the other, its mass is gravitationally "attracted" by these wave ripples. As soon as the particle enters one of these ripples, the particle follows the intensified light until it reaches its final destination point on the screen in an orderly "wave-like" fashion.

With this simple classical "electromagnetic" explanation out of the way, there is absolutely no need to assume a particle had mysteriously passed through both slits at the same time as the mathematics of quantum mechanics might imply. Nor do we have to assume the particle has a mind of its own in order to know how to behave in a wave-like manner. The explanation tells us it is irrelevant which slit the particle entered or what the particle might "want" to do should the opportunity present itself within the experiment. It just happens naturally through a simple "cause-and-effect" relationship, which can be described by classical science. Physicists just have to remember that radiation has a gravitational field of its own (if they wish to retain this field as an independent and real force of nature), and must use their imaginations to see how the particles would behave classically under the influence of this gravitational field in radiation. Alternatively, under the laws of electromagnetism and the Unified Field Theory, there is no gravitational field. Therefore, imagine, from a purely electromagnetic perspective, radiation pushing (or guiding) matter toward a certain destiny.

This is the fundamental problem of the modified Thomas Young experiment. Removing the observable monochromatic light source and throwing a bunch of particles at the double-slit experiment does not necessary imply the particles are free from all interference by radiation. We still have the natural universal background radiation to contend with, which is penetrating the experiment all the time. Thus, physicists need to ask themselves, "What exactly is this radiation doing to the experiment?"

Come to think of it, what would happen if we remove the radiation in the modified Thomas Young double-slit experiment.? Would this remove the intensified light ripples in space? And if so, would the electrons passing through the slits behave exactly like particles by bunching up as two distinct bands in front of the two slits (or whatever number of slits are present)? Irrespective of the size or speed of the electrons or any other quantum particle you throw at this experiment, no solid matter should ever show wave-like properties in a universe free of all radiation.

What Role Does Probability Play in the Quantum World?

However, as we have it today in the world of quantum theory, we have to use simplified mathematics to apply a probability game to the Universe. This is perfectly fine if a scientist needs a solution and he doesn't want to solve a complex equation with many variables and other factors. It is okay to keep things simple, so long as we understand what we are doing and can keep in mind the real picture that is taking place in reality, which is the universal background radiation will naturally cause the particles to move around in all directions to give a sense of randomness. Then this randomness of the quantum world can be represented by the laws of probability in a simplistic way to help determine how likely a particle is to be at a certain place and at what speed. Likewise, particles could potentially be guided by radiation to move in specific directions, or get held in specific regions. Again this can be simplified mathematically using the laws of probability. But does this mean Newtonian physics can never come up with the same answer? Despite the fact that one never knows precisely the position and speed of a particle simultaneously as one can in Newtonian physics, there is still hope for Newtonian physics to make its comeback and settle certain long-standing issues that Einstein had with quantum theory. Sure, the mere act of observing a quantum particle using radiation disturbs it so much that we cannot determine the two things we need to define an infinitely thin path of where the particle is going and how fast. Despite this limitation, it does not mean particles do not move in specific paths in space. They do. They can still follow the

154

classical laws of Newtonian physics. It is just that if we attempt to apply the Newtonian equations to the quantum world, they will be too complex and with things moving too fast. At the same time, running the equations in a computer simulation over time may help to reduce the time to calculate things and do it en masse for the number of equations needed to get meaningful answers, but working out what one quantum particle might do and proving it in reality will be too difficult. Instead we have to look at a group of particles and determine how likely they will exist in a certain place and its speed. Even though we can never prove the exact path taken by a quantum particle and instead use a simplistic quantum equation to calculate a probability number of how likely the particle will exist at a certain place and at what speed after a period of time, it does not mean Newtonian physics cannot be applied to the quantum world to come up with the same probability answer. It can at the most fundamental level. It all depends on how many equations one is prepared to solve and how fast so that one can see reality get played out to help us indirectly see what the particles are doing.

Therefore, we should expect these electrons to move in a random, as well as particle-like or wave-like way. It is natural. But it is not because quantum particles have minds of their own to know when to show particle- or wave-like behaviors or to choose where they like to be found and at what speed in a certain place and time—nor is it because God[5] is playing dice. Neither is the fact that the particles are splitting apart as they move through the slits, as the mathematics of the quantum theory suggests. That is nonsense. The particles remain as particles throughout the entire path to whatever destination they eventually will end up. And the thing that is guiding the particles to their final destination is radiation. Yes, the ever so pervasive substance of the Universe. It comes from different directions at different frequencies. It gets filtered through the narrow slits and bends around solid matter through the process of diffraction at different angles. The radiation merges and combines in a constructive and destructive interference manner to form intensified and stationary light ripples in space that help to push quantum particles to specific regions. Without

---

5   In Chapter 8, science will define God in a way that allows light and God to be seen as synonymous terms. Thus, it may be totally scientific to say that God is influencing the particles. We will have more to say about this God issue later.

these slits, the particles naturally move at different speeds and in different directions. This is how the randomness comes into the picture. It is only in the event that radiation is constructively and destructively interfered to create these invisible and interesting stationary light ripples in space do we find the particles influenced by the ripples and get pushed in an orderly pattern to create the wave-like effect.

So when physicists talk about probability waves in the context of quantum mechanics and the equations that are applied, they are referring to the radiation of the Universe playing its role in the quantum world. The purely mathematical wave does have a real-life equivalent wave. According to the Unified Field Theory, this is called "radiation".

There has been a great debate in quantum mechanics on the issue of whether probability waves are real waves. As Nazim Bouatta, a theoretical physicist at the University of Cambridge, has noted:

> "The question was, we have this wave function [i.e., Schrödinger's equation], but are we really thinking that there are waves propagating in space and time? De Broglie, Schrödinger and Einstein were trying to provide a realistic account, that it's like a light wave, for example, propagating in a vacuum. But [the physicists], Wolfgang Pauli, Werner Heisenberg and Niels Bohr were against this realistic picture. For them the wave function was only a tool for computing probabilities."[6]

EINSTEIN ATTACKS QUANTUM THEORY

Scientist and Two Colleagues Find It Is Not 'Complete' Even Though 'Correct.'

SEE FULLER ONE POSSIBLE

Believe a Whole Description of 'the Physical Reality' Can Be Provided Eventually.

That's fine to have a "tool" to use in order to determine the probabilities associated with quantum particles. You can use as many mathematical tools as necessary to find an answer. The real problem here is not relating the mathematics to reality. What exactly are physicists doing when they apply the mathematics? What is actually happening in the real Universe we live in to influence

---

6   Quote from http://plus.maths.org/content/schrodinger-1.

and guide quantum particles in a way that increase their probability in being at a certain location and a certain speed?

That is what's missing in quantum theory, and is the reason why Einstein never believed God plays with dice in the Universe. There had to be a "cause and effect" explanation. No voodoo magic or a mysterious God. Just plain physics with particles colliding and radiation guiding them.

Sure, we can never actually see with our eyes how these particles move. That is why we have the Heisenberg Uncertainty Principle established in quantum theory. This principle acknowledges the limitations in physically observing the quantum world and getting the necessary information to determine exactly what the particles are doing. Yet this limitation should not cause us to shut down our imaginations, use simplistic mathematics to calculate the probabilies of the particles to be at a certain place in time and at what speed, and accept whatever bizarre solutions the mathematics tell us. We can use our minds to understand what is likely happening. From there we see the real wave of nature that is causing particles to behave as they do—specifically, it is radiation. Once we see the "cause and effect" world being played out by radiation on the particles, it should be possible for computer simulations to confirm in our minds what we can visualize is the answer.

This is something the physicists will have to consider carefully in the light of what we know about the Unified Field Theory. Either we assume the quantum world is ruled by mathematics (i.e., probability waves) because we cannot observe what is going on[7]. Or we use our imagination and enact classical science by using the gravitational field in the radiation to show how one can increase the probability of where particles are likely to be found and at what speed. As we know, particles are gravitationally attracted to each other and to regions that generate this field. Makes perfect sense to imagine the gravitational field of the radiation to pull in particles and push them around. That is how one increases the probability of particles entering certain regions of space. Or else one can accept the extraordinarily bizarre implications of mathematics in all its intricate details. Things like particles being able to split apart as they pass through two slits and come together again. Huh?

---

7   While accepting the paradoxical behaviors that quantum particles can display under certain conditions.

What on Earth were physicists drinking at the time to come up with this ludicrous picture?

Here's an example.

According to quantum mechanics, we are told, in a mathematical sense, how a change in state of one quantum particle can instantaneously influence the state of another quantum particle positioned anywhere in the Universe and vice versa. Known as "quantum entanglement", the mathematics of quantum mechanics suggests that whatever it is that's traveling between the two quantum particles to influence each other (the mathematics does not attempt to explain it, other than as a probability wave) instantly has to be traveling beyond the speed of light. However, light only travels at 300,000 kilometers per second. Surely it can't be light. It must be something more exotic traveling between the quantum particles—something we don't see in reality. Why?

The fact that the mathematical solution exists is enough to make some scientists think it could be a reality. And, because light is limited to a certain speed, physicists have to consider the possibility of a new and exotic particle yet to be observed in the visible universe that can achieve this remarkable feat.

Now really?

What if physicists have overlooked the fact that a universe created by mathematics is one already devoid of universal background radiation? Could this be an important point to consider? Take the radiation out of the mathematical universe created by Schrödinger's oversimplified equation, and quantum particles can still be shown to perform all the same classical behaviors we expect in Newtonian physics and relativity. It is just that we now have to remember that the speed of light is infinite in a perfect vacuum. Put the universal background radiation in, and the classical laws of physics still apply, so long as we realize that radiation travels at the familiar 300,000 kilometers per second. If we don't take into account this radiation, the reason why the state of one quantum particle can influence another at any distance, even infinite range, is because radiation acting as the messenger in delivering the information to another quantum particle is traveling at infinite speed. Not only that, but it also stretches out to infinite distances in a perfect radiation-less universe described by the humble mathematics. Thus, radiation has the required electromagnetic tentacles to reach out and affect any quantum particle at any distance to change its state. This is what happens when Schrödinger's equation is oversimplified.

So, far from being bizarre, quantum entanglement has a natural classical Newtonian explanation. No need for an exotic particle to explain the so-called impossible speeds needed to achieve this supposedly miraculous effect. Nor do we have to resort to God for an explanation.

If everything said here is true, then there is no need to consider exotic particles/energy or new forces of nature to explain what is happening. The seemingly bizarre nature of the quantum world through things like quantum entanglement can have a classical explanation. All we need to do is be aware of the influence that light and its own gravitational field has on quantum particles and how the lack of radiation permeating space can make light travel at much faster speeds to support the mathematics. It is here, in a mathematical universe devoid of radiation and relying on the laws of probability for answers, that all the bizarre predictions from quantum mechanics will come.

Quantum Mechanics to Match Reality Needs to Include Radiation

So where do we go from here?

Well, if we want to continue relying on quantum mechanics in its current form for an answer, it must be tempered by reality because of the presence of *universal background radiation*. That's the only difference. The Unified Field Theory is meant to account for this radiation within its own mathematical universe. The conditions set up to solve quantum problems using Schrödinger's equation are simply not set up to include this radiation and the gravitational field it generates to influence quantum particles. This is why mathematics in quantum mechanics do not relate well to reality: mathematics must take into account the real Universe we live in, which happens to be filled with the universal background radiation.

It is only when we realize the impact of this radiation in the quantum world and the presence of the gravitational field in the radiation that we can appreciate how all the scientists are right in their views, including Einstein himself. His unified field equations are essentially quantum field equations. It is just that, in order to be precise in the paths taken by so many quantum particles, the equations have to be naturally complex. Or we can simplify the situation using probabilities thanks to Schrödinger's equation. Which would you prefer? Naturally, nearly all physicists prefer the latter. Seriously who wants to spend years solving the unified field equations for a quantum problem involving countless numbers of particles? Too much work. What the unified field equations should be able to do is provide the same answer, but without all the bizarre results derived from the oversimplified quantum equation. The unified field equations already take into account the radiation in the Universe. That is why they are complex to solve.

At some point, Schrödinger's equation of quantum mechanics in its current form and the unified field equations of Einstein's Unified Field Theory will eventually be seen as one and the same theory.

CHAPTER 6

How Old and Big Is the Universe?

*Two things are infinite: the Universe and human stupidity; and I
am not yet completely sure about the Universe.*

—Albert Einstein[1]

A T THE present time, scientists have accepted the view that the
universe is finite and expanding and has an age of between 10
and 20 billion years (the best estimate so far is 13.7 billion
years[2]).

Physicist Dr. James Trefil summed up the prevalent scientific view
on the universe when he said:

"The universe is expanding. Every way you look from the Earth, you see galaxies receding, with distant galaxies receding more rapidly than those closer in. Now, if we imagine tracing this observed expansion backward in time, and if we use the simplest possible logic, we find that the universe in earlier times was denser than it is today. And at some time between 10 and 20 billion years ago, all the matter in the cosmos was packed into a single, infinitely dense point. Such a point— similar to what may lie at the heart of a black hole—is called a singularity."

Let us review the evidence supporting the "finite and expanding universe" view and determine if there could be an alternative explanation based on what we know about radiation.

Currently Accepted View of the Visible Universe

Redshifting effect of light

In 1912, Dr. Vesto M. Slipher became the first scientist to observe a redshift in the light of distant galaxies at the Lowell Observatory in Flagstaff, Arizona, USA. This redshift effect is more prominent with distant galaxies than with those closer to Earth. Why the redshift occurred and would increase with greater distances remained a mystery at the time.

It wasn't until 1929 that American astronomer Dr. Edwin Powell Hubble (1889–1953) made the first and only official scientific explanation for why this redshifting effect was being observed. With no other competing theory to consider in his study of the redshifting effect problem, he promulgated the view that the visible universe is probably finite and expanding. He did this after applying the *Doppler effect* to the stretched-out light that the galaxies he had observed were emitting, using the 100-inch Mount Wilson telescope in California, USA.

The Doppler effect is said to be a measure of the shift in the frequency of the light that objects moving toward or away from an observer emit. If the frequency of the light shifts toward the red end of the spectrum—known to physicists as a redshift—it is interpreted as the light source receding from us. A shift toward the blue end of the spectrum—known as a blueshift—means the light source is approaching us.

Dr Edwin P. Hubble

After measuring the degree of the redshift in the frequency of light received from distant galaxies, Hubble noticed that if a galaxy is calculated to be about twice as distant as another, its speed of recession will appear to be twice as great. It is as if the Doppler theory is telling us that all of the galaxies are racing away from the epicenter, which, extraordinarily, happens to be where our planet is located.

Spiral galaxy NGC1376. With the exception of the closest galaxy to us—the Andromeda Galaxy—this object and all other galaxies are said to be racing away from us as if our visible universe started from a Big Bang. Is this true? Or could there be another explanation?

Of course, such frightening images from the past of us being at the center of the universe, as if God has favored us over all things, did not sit well for the scientists. It took a while, but eventually another interpretation arrived in the nick of time to give Hubble's theory slightly more credibility.

The new interpretation would arrive after scientists took a closer look at the General Theory of Relativity. Now instead of an epicenter to be at or near our planet, imagine we are sitting on the surface of a balloon (the familiar two-dimensional fabric explanation as we have discussed in Chapter 2). As the balloon expands, the galaxies would appear to be racing away from us without consideration that the Earth is at the epicenter.

In the meantime, Hubble thought that if his interpretation of the redshifting effect is true, there must have been a time when all of the galaxies were much closer together.

Belgian astronomer and priest Georges Lemaître provided the clue Hubble needed. In 1927, Lemaître suggested that "all the matter in the universe was once condensed into one huge mass that became unstable and exploded".

When Hubble heard of Lemaître's idea, he accepted it. He then published the idea together with his results in a scientific journal. Today, the theory of the "Big Bang", as it is called (and in the absence of any alternative explanation for the redshifting effect), is one of the most hotly debated theories of modern science.

Upon hearing about Hubble's work, Einstein could not think of an alternative explanation for the redshifting effect. He only just published his Unified Field Theory. Without enough time to solve the equations to see if a solution would validate or contradict Hubble's work, he thought Hubble's idea was plausible. So, in response, he added a cosmological constant to his gravitational field equations to show this stretching of "space-time" to be in keeping with Hubble's official explanation.

But was this a wise move from Einstein?

Alright then. If the Big Bang theory is true, what additional evidence can we find to support it?

Supporting the Big Bang theory

In 1948, Ralph Alpher, a student who worked with the exceptionally talented nuclear physicist George Gamow (1904–1968), as part of his PhD thesis, provided what could be supporting evidence. As Professor Hawking said:

> "In this paper they [Alpher and Gamow] made the remarkable prediction that radiation (in the form of photons) from the very hot early stages of the universe should still be around today, but with its temperature reduced to only a few degrees above absolute zero (-273°C)."[3]

---

3   Hawking 1988, p.124.

In other words, if a Big Bang did occur, the universe should have been flooded with intense gamma and X-ray radiation shortly after the explosion and would have redshifted into the microwave region after billions of years of interaction with solid ordinary matter (presumably without any similar interaction with the radiation itself).[4]

Direct observations in support of the theory came in 1965 when two American scientists—Dr. Arno Penzias and Dr. Robert Wilson—from Bell Telephone Laboratories found, by accident, the universal background microwave radiation of about 3 degrees above absolute zero Kelvin, a discovery for which they won a Nobel Prize in 1978. Although at the time they could not account for the source of this intriguing radiation, astronomers now interpret this as the "cosmic echo" of the Big Bang.

Church supports Big Bang theory

Sufficiently convinced by these seemingly solid scientific arguments, available observations, and the apparently correct interpretations made by scientists of the latest observations in relation to what the universe has allegedly done in the past (i.e., it started from a point in space) and is still doing today (i.e., expanding), and seeing how this would support the idea of God creating the universe at the beginning of time and space and an even larger Universe outside of it, Pope Pius XII of the Roman Catholic Church made the idea an official position of the Vatican in 1951:

> "In the beginning, according to the Book of Genesis, God created the heaven and the Earth. In the beginning, according to 20th-century astronomers, there was a Big Bang. To show that the two views—religious and scientific—were totally compatible, Pope Pius XII adopted the big bang theory of the universe as the official policy of the Roman Catholic Church in 1951."[5]

---

4 It should be remembered that Einstein also wanted to consider radiation as ordinary matter as well, according to the Unified Field Theory. This will have important implications later in this chapter. Furthermore, this redshifting of high-frequency radiation at the beginning of the universe should have been accelerated by the presence of other radiation in space. So does this mean the age of the universe is much younger than the predicted 13.7 billion years as scientists currently believe?

5 Stemman 1991, p. 75.

As if the Big Bang theory could use extra support from religious leaders to make it seem more correct, we also find numerous scientists who have not yet been deterred from supporting the Big Bang theory as well.

So, what are we to make of these observations? Indeed, what is the visible universe going to do in the future? Will it expand further and later slow down at some point in time? Or will it keep on expanding forever? Or have scientists made an incorrect interpretation on the redshifting observation which could see the universe in a steady state or possibly contracting?

The answer to the future of our universe, assuming the Big Bang theory is correct, will depend on how much matter exists in the universe.

Mystery of dark matter

To determine the total amount of matter in the universe, scientists made further prudent observations.

In the early 1970s, Dr. Vera Cooper Rubin from the Carnegie Institution made it her job to record time measurements of the rotational speed of stars orbiting the great black holes lying at the centers of every galaxy we can observe in the universe (including our own galaxy, the Milky Way).

As Rubin said:

> "When I started working here, I looked for a problem I could do at my own pace. I had four children and a supportive husband but traveling was more difficult to go to observatories out west which was necessary."[6]

At the time, black holes were all the rage for the astronomers. Not so for Dr. Rubin, who focused her attention on galaxies and, specifically, the speed of rotation of stars at various distances from their galactic centers. "If we plot the velocity of the planets as a function of distance from the Sun," Dr. Rubin noted, "you can see that Mercury orbits much more rapidly than Pluto."

---

6   All quotes from Dr. Rubin obtained from the documentary titled *Most of Our Universe is Missing* televised on SBS on July 27, 2008.

Dr. Vera Rubin draws on a blackboard the expected graph of how the gravitational field weakens the further you move out from our solar system.

The "rotation curve" as this graph is called, is said to be an embodiment of Newton's law of universal gravitation. It shows that the farther away the planet is from the sun, the weaker the gravitational field gets. Nothing unusual here, Rubin thought.

Next, she began looking at the stars in the Andromeda galaxy, the closest to our Milky Way. In particular, she specifically focused on those stars sauntering around the outskirts of the galaxy. As Dr. Rubin explained:

> "I picked the outsides of galaxies because that was the problem no one...was studying and it was a problem that I had been interested in for quite a while."

Since the stars were being held by the gravitational field of a black hole lying at the center of the galaxy, she was curious to see the kind of rotation curve generated by these stars. When she plotted the velocities of the stars against distance from the galactic center, she noticed they were not slowing down the farther away they were from the center. Dr. Rubin said:

> "I came out with sets of numbers and I plotted them on pieces of paper and I discovered that the stars as you went

further and further out did not slow down, they were moving just as fast as the stars near the center. And that was a surprise. And a surprise that had to be explained."

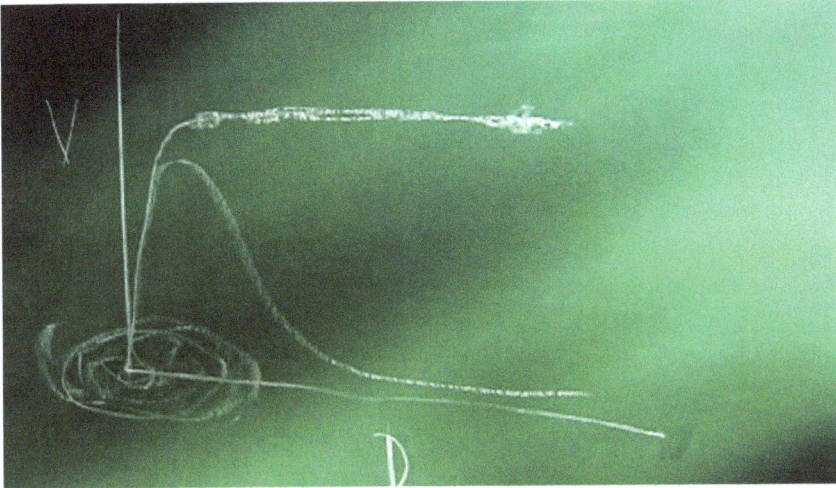

The bottom line drawn by Dr. Rubin is the expected graph if the stars had followed Newton's law of gravitation the further they were positioned from the center of a galaxy. The top line represents the actual results obtained, showing an unexplained anomaly.

Dr. Rubin had uncovered something very odd. Unless the stars were defying the laws of gravity, she had to consider an additional gravitational field from some mysterious extra mass being added over the entire length of the galaxy to make the stars rotate at the same velocities. Below shows the actual graph Dr. Rubin had plotted from her original observations.

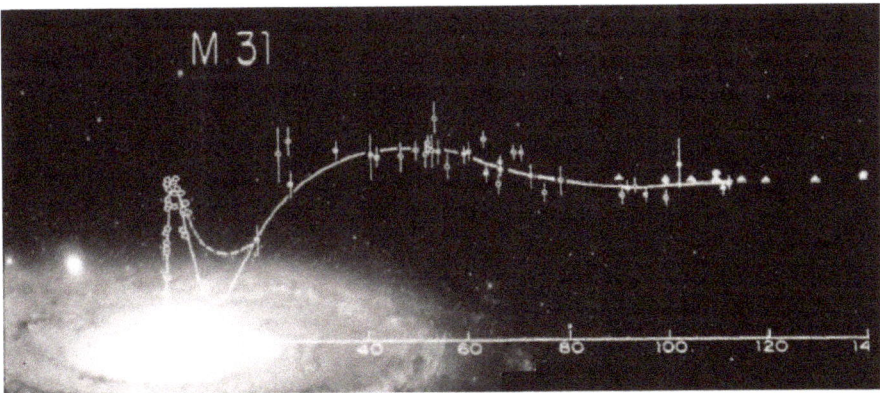

Whatever this mysterious invisible mass may be, hidden between the stars to create this intriguing pattern, scientists have been calling it *dark matter*.

But that's not the end of the story.

Mystery of dark energy

The next piece in the great puzzle of the universe came by way of the interesting work of astronomer Professor Brian Schmidt of the Mount Stromlo Observatory; physicist and astronomer Professor Adam Reiss at the Space Telescope Science Institute in Baltimore; and Professor Saul Perlmutter from the University of California, who won the prestigious Shaw Prize for astronomy in 2006 after noticing the expansion rate of the universe was apparently accelerating in direct contradiction to the commonly held scientific view that gravity should be slowing the expansion of the visible universe. It seems the galaxies are accelerating away from the Earth more quickly than expected over time.

As Professor Perlmutter explained:

> "At the time, it was pretty clear to us that the universe should be slowing down because of gravity. All the material in the universe attract each other because of gravity and you would expect that that would slow the expansion of the universe. What we didn't know at that stage was whether or not it would be slowing down enough and someday come to a halt and then collapse and that was the fundamental objective."[7]

Professor Perlmutter and his colleagues measured how fast exploding stars, known as supernovas, in distant and nearby galaxies were receding from us by analyzing the light in these stars. Professor Perlmutter said:

> "What we saw was that the universe apparently is speeding up. And that was the surprise. We did not expect the universe to be speeding up in its expansion."[8]

---

7   Documentary titled *Most of Our Universe is Missing* televised on SBS July 27, 2008.
8   *Most of Our Universe is Missing.*

It is as if some kind of *dark energy* (or force) is repelling or breaking apart our visible universe and counteracting the force of the gravitational field from all matter in the universe.

If the interpretation of the observations is correct, scientists are thinking this repulsion force will cause our visible universe to disappear in the not-too-distant future (well, at least another 13.7 billion years from now) as all galaxies outside the Milky Way eventually disappear from view.[9]

Still, this brings us back to the disturbing idea of why we are so lucky to be experiencing the pleasure of seeing the stars and galaxies in their present distances to allow us to observe them when, technically, we could be living at a time when there are no galaxies, or perhaps nearer to the beginning of the universe when other galaxies were much closer or still forming into galaxies. Are we to infer from this interpretation from the scientists that we are at the center of the universe in terms of time itself? Already the bells of concern are ringing again.

With so much riding on this "expanding universe" idea, how are we going to explain this mysterious dark matter and dark energy? Surely, there must be a good explanation.

Percentage of matter in the universe

If the universe is said to be expanding and galaxies are racing away from each other, what percentage of matter (or energy) exists in the universe?

Assuming dark energy and dark matter exist, the Big Bang theory should predict the total mass in the universe today and with it the likely percentage of the mass due to dark energy and dark matter, as well as the visible matter from stars, planets, and gaseous and dusty nebulas. And if this is true, do the predictions correlate to observations of the universe?

After making a few simple calculations, scientists estimate that we now have between 68 to 74 percent of the universe is filled with some kind of dark energy creating this apparent expansion. This leaves us with 22 to 27 percent as the dark matter that provides the extra

---

9   With the exception of the Andromeda Galaxy, which will merge with the Milky Way in 5.1 billion years from now.

gravitational force for making stars rotate around galaxies at the same velocities, and less than 10 percent visible matter (best estimate is probably 4 percent) composed of atoms (including the matter that is unable to reflect or emit light).

Supporting this 4 percent figure for visible matter in the universe is the work of Professor Carlos S. Frenk of Durham University. Frenk conducted a survey of the night sky by counting the number of stars. Assuming these stars contained additional matter by way of planets and moons, and if he included other solid matter that did not reflect light or generate light of their own (such as the dust we see blocking our view of the Milky Way center), he estimated that approximately 4 percent of the universe contains this solid matter. Therefore, he reasoned the universe must have 96 percent dark energy and dark matter.

So far, the mathematics and careful analysis of the visible universe seems to allow for the existence of dark matter and dark energy or, at the very least, help support the observations by way of the matter visible to our eyes while maintaining the view that the universe must be expanding. The only problem is, what is this dark energy and dark matter in reality?

In the case of the mysterious dark matter, scientists are postulating a new, exotic particle permeating the visible universe called a *neutralino* as the thing responsible for this. As Professor Tim Sumner of Imperial College London said:

> "The form of dark matter that we're trying to look for is a new family of particles. We have good reasons to believe these particles exist because they are required by other theories within physics."[10]

As for whatever is causing this mysterious dark energy to expand the universe, your guess is as good as any so far.

Ripples in the Universal Background Radiation

With everything seemingly fitting in with observations to support the Big Bang theory, scientists believe they have found the final piece of

---

10   *Most of Our Universe is Missing.*

Artist impression of the Wilkinson Microwave Anisotropy Probe (WMAP) launched in June 2001.

evidence to end all speculation about what is happening with our universe.

In the 1990s, scientists believed they had found support for the Big Bang theory after detecting extremely faint ripples in the universal background radiation (especially apparent in the microwave region) using NASA's Cosmic Background Explorer (COBE). Following a 2-year pass to gather the light from the farthest regions of the visible universe, the result (together with the Milky Way shown in the middle) can be seen in the top image on the following page.

Upon closer analysis, scientists interpreted these ripples as the primordial soup of photons, atomic nuclei, and the mysterious dark matter, which they believe is evidence of the Big Bang and thus the finite nature of the universe. Why? Because the ripples represent extra matter by way of gas getting ready to combine under gravity to form new stars and galaxies.

However, with further refinement of their instruments and with longer time frames to gather the light, scientists are discovering how quickly the ripples are getting smaller and more resolved. For example, the middle image (without the Milky Way) was obtained after 4 years were spent gathering more light from what scientists believe is the edge of the visible universe.

The ripples are now looking like blobs, yet scientists are still happy to interpret these blobs as matter ready to form new stars and galaxies (in which case we might as well believe in Santa Claus, the Tooth Fairy, and the Easter Bunny). Thus, what we are observing here is the residual image of a time when the universe was a mere 360,000 to 400,000 years old.

Still, technology has advanced and become more sensitive to help refine these "blobs".

Launched on June 30, 2001, the Wilkinson Microwave Anisotropy Probe (WMAP) has provided more accurate and detailed observations of the universal background radiation. This time, after undergoing a 7-year pass, the blobs have made way for what look more like points of light emerging from the background noise as the bottom image shows. The temperature range for this image is between plus and minus 200 microkelvin.

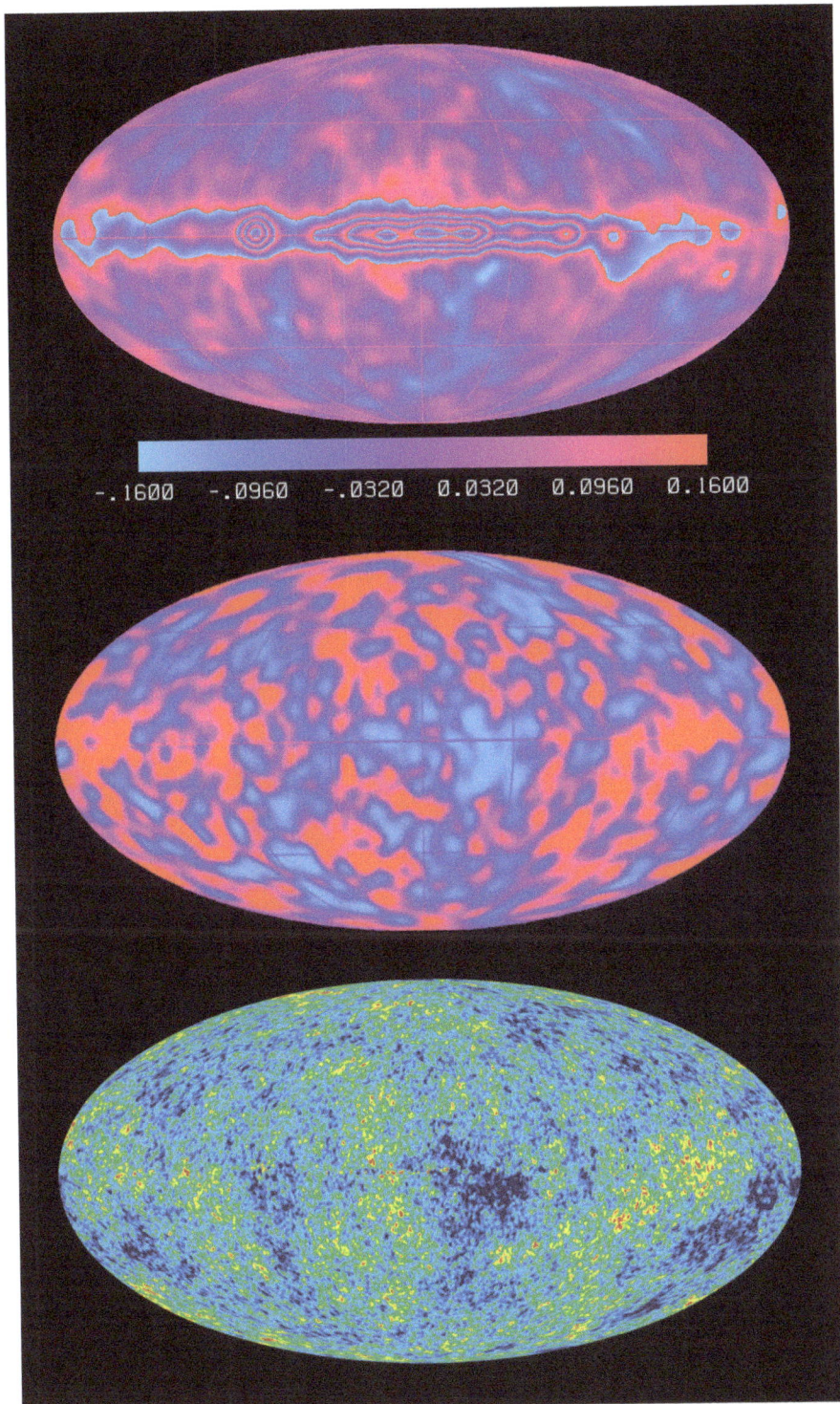

Again, scientists still do not expect to find any structures that are reminiscent of galaxies or superclusters of galaxies hidden inside any of these bright red and yellow spots. Scientists believe that these mysterious spots are just a bunch of empty gas, ready to form new galaxies and stars.

With the way in which things are progressing at the moment, perhaps we should believe in Father Christmas. Or, is there another explanation for these spots of light?

So, what's the counterargument against the possibility of more distant galaxies emitting extra radiation within these blobs of light in the WMAP chart?

"Because otherwise," as scientists would argue, "an infinite Universe would imply absolutely no ripples in the background microwave radiation because galaxies would exist everywhere. And the Doppler effect and the gravitational field equations tell us that the amount of energy in the radiation at every frequency would remain the same in every direction."

This is how the chart should look if we were living in an infinite Universe[11]:

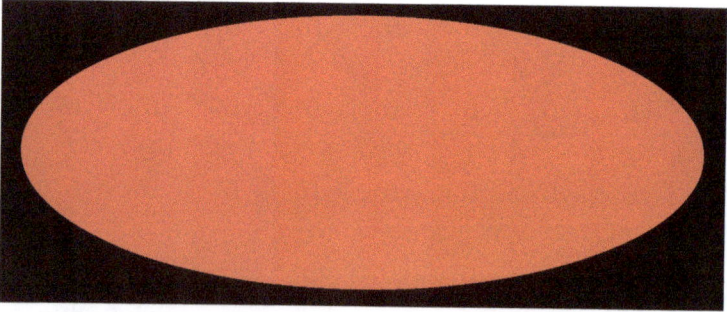

In an infinite Universe, the temperature would have to be constant throughout. The assumption for this view is that radiation does not lose its energy by interacting with itself; only solid, visible matter interacting with radiation will cause this "redshifting effect" in the radiation.

An interesting argument. The only tiny niggling issue is that all of this assumes that radiation traveling through space would not undergo any form of natural redshifting of its own.

---

11   We capitalize the word here because it is the ultimate. The lower-case form is used to represent a portion of the Universe (e.g., the visible universe, considering that there has to be an invisible universe outside of it) or if evidence for multiple universes exists.

The Unified Field Explanation

Yet the Unified Field Theory tells us that we must treat radiation of any frequency as ordinary matter. And what does ordinary matter do when it interacts with other ordinary matter? A loss of energy in the radiation (i.e., redshifting) takes place as it travels through radiation-filled space. It happens all of the time, and is perfectly natural and expected. It is an energy loss that need not affect all objects in the visible universe. In particular, in an infinite universe, other radiation can arrive to counteract any expansion. Thus, the universe may not be expanding. Furthermore, the radiation generates a gravitational field. How interesting. Could this radiation help to explain the dark matter in a galaxy?

The Compton effect

The experimental evidence in support of this photon-to-photon collision leading to a redshifting effect in radiation comes from British physicist Dr. A. H. Compton (1892–1962). In 1921, Compton became the first to study the redshifting effect of the frequency of X-rays scattered at various angles by the electrons in a graphite block. He noticed that when X-rays of wavelength λ collided with an electron, the radiation deflected by the electron had a wavelength differing from the original by an amount $\Delta\lambda$ called the "Compton shift". This shift increased as the angle of deflection increased (thus indicating a more direct hit).

Also, this shift became redshifted (not blueshifted).

To explain this observation, Compton had to assume that X-rays behaved like a stream of particles (i.e., ordinary matter). Whenever the X-rays collided with an electron, some of the energy in the X-rays appeared to be transferred to the electron, causing the electron to "recoil" in one direction and the photons to go in another with a slightly lower frequency (i.e., less energy).

If this is true, then surely the Unified Field Theory is telling us that radiation against radiation should achieve the same results. Why? Because radiation is no different from, say, an electron or a tennis ball. Radiation has to lose some energy in its collision with other radiation and solid matter. The effect may be less noticeable at low frequencies, but it does exist.

A question was put to NASA scientist Michael Loewenstein about the speed of light in different energy densities of the mass-energy of space, as well as the red-shifting effect of light colliding with other light:

> "Scientists say the speed of light can be made to slow down when light passes through a dense transparent material. For example, light moves slowly through a diamond, than in glass. Generally this is because the electrons of atoms are temporarily excited by the energy of the light and there is a time delay before the energy is released again. In the vacuum of space, we have some electrons, protons and free moving positively charged particles such as hydrogen and helium. However, the most abundant particle is the photon. Photons are said to be ordinary particles as Einstein believed and can collide with each other, causing light to red-shift, bend etc. Speed of light might be 300,000km/s in this vacuum, but what's the speed in a perfect vacuum containing no radiation?"

Loewenstein responded by saying:

> "The constant that is usually referred to as the speed of light *is* the speed of light in a vacuum. As you note, the universe is not empty, but the chances of a photon colliding with an atom along the way is exceedingly small. Photons are much more numerous (by more than a factor of a billion), but the probability of interaction between two photons is very small unless the photon energies are much higher than is the case for those contributing to the cosmic microwave photon background. For that reason the actual speed of light in the universe is indistinguishable from that in a [true] vacuum."

Really, is it so negligible? Or, is this just another way for the scientist to say, "We don't think so, but we are not sure. And we haven't tried performing an experiment or doing any kind of calculations yet to find out."

Well, it seems that some other scientists have carried out an experiment using a metamaterial containing a "nano-scale structure that guides light waves through the metal-coated glass...with a refractive index below 0.1". A low refractive index is very much like having a low mass-energy density of space. The higher the refractive index, the denser it gets. For example, the radiation-filled vacuum of space is set to 1.0. Air is a higher mass-energy density environment having a refractive index of 1.000277. Water is denser with a refractive index of 1.333333. And a diamond has a refractive index of 2.417, showing how much more denser this material is. So, if you reduce the refractive index to below 1.0, you are effectively approaching the true vacuum containing absolutely no radiation. Well, in this very low mass-energy density environment having a refractive index below 1.0, Albert Polman at the FOM Institute AMOLF in Amsterdam, the Netherlands, who has studied the metamaterial, doesn't quite see the change in the speed of light as negligible at different mass-energy densities when he says light passes through the material at an almost infinite speed. There is a big difference between 300,000 km/s and infinite speed. This might be a good opportunity to refer back to Chapter 2 and to check out the original article published in *New Scientist* on January 9, 2013, for details about this experiment.

Loewenstein is also quick to be dismissive of the alternative theory for the redshifting effect mainly because at low frequencies (or photon energies), the probability of interaction is thought to be exceedingly small unless the photon energies are much higher. At high frequencies (or high photon energies), the effects of such interactions are observable (e.g., Brodsky, Stanley J., "Photon-Photon Collisions—Past and Future"[12]). However, the probability of such interactions occurring in the natural "vacuum of space" (which we know is not a true vacuum when the radiation exists) above the Earth's atmosphere is low due to the number of these high-energy photons in the cosmic microwave photon background, whereas the interactions at low frequencies are much more numerous, but the redshifting effect is thought to be negligible. Even if the volume of space for photons to travel through is very large, such as the distances between galaxies, the cumulative effect of the light that redshifts with each successful interaction apparently

---

12   Article published in November 2005. SLAC-PUB-11581, downloadable from
https://www.slac.stanford.edu/cgi-wrap /getdoc/slac-pub-11581.pdf.

remains negligible by the time light from a distant galaxy reaches our instruments.

If Loewenstein was incorrect in his assessment of the speed of light in a true vacuum, would it seem reasonable to think that perhaps he might also be a little off the mark with his assessment of the redshifting effect of radiation at frequencies within the visible spectrum?

Something is telling us that this alternative theory for the redshifting effect based on photon-to-photon collision should be checked more closely. One can't simply brush it off by saying the redshifting effect of radiation against radiation is negligible at any distance at all frequencies. Surely, some measurable difference has to exist. The question is, How much of a difference are we talking about here? A lot? A little? Or enough to mask whatever redshifting might exist based on the Doppler theory and thus make it impossible to know for sure what the galaxies are doing?

In fact, as we will discover later in this chapter, observing the edge of the visible universe will reveal a paradox. It means that two opposing, but equally valid answers must exist for what is happening in the universe. One of them is the Big Bang theory with its suggestion of a finite but expanding universe. The other must be that the visible universe could already be infinite and be called the Universe, and no further expansion is necessary. Everything would be in a steady-state. Whatever explanation we give for the redshifting effect, it does require the scientific community to provide an alternative explanation. Scientists cannot accept just one theory to support one interpretation and not have an opposing theory that can explain the same evidence at the same time.

It seems reasonable at this stage to consider the photon-to-photon theory for the redshifting effect.

Therefore, the assumption that radiation redshifts only when it interacts with solid matter is not entirely correct. We must include other radiation in the mix. Furthermore, any spots of raised temperature readings that we can observe in the cosmic background radiation could easily represent more distant and unseen galaxies and superclusters of galaxies.

As for the assumption that any kind of redshifting in radiation would cause objects to move away from each other is not true. In an infinite Universe, radiation from the edge of the visible universe arrives in roughly equals amounts and covering the same frequencies. This means radiation can exert an equal force from all directions. Add to this the natural energy loss and it eventually becomes much harder to detect whether a collision did take place at sufficient distances. It is also less likely for objects to move away from the collision (or explosion if you are thinking about the Big Bang). Just like throwing a stone into the middle of a large pond and watching the energy dissipate to the surrounding waters, it hardly affects the positions of the boats sitting far enough away from the source of the disturbance. The same is true of radiation. In an infinite universe and with radiation redshifting, galaxies do not need to move away from each other at sufficient distances. Galaxies can still move in any direction and at whatever speed they do. The problem for the scientists is, how do you prove it? You just can't tell when you approach the edge of the visible universe. Galaxies have to be close enough for scientists to potentially discern the difference in redshifting due to the Doppler effect compared with the natural energy loss in space from the countless photon-to-photon collisions that occur. In the case of the Andromeda Galaxy, it is close enough and fast enough for scientists to detect a blueshift in the light to show that the galaxy is approaching the Milky Way. But for other galaxies in the Universe, they could very well be too far away to give any meaningful measure of the redshifting effect based on a possible receding of the galaxies using the Doppler theory. Indeed, why would the Andromeda Galaxy be an exception when countless other galaxies are moving away from the Milky Way? Surely there must be plenty of other galaxies that could be heading our way.

The collisions that take place between radiation and the rest of the radiation-filled space must show some level of redshifting given enough distance. This occurs more so in the higher frequencies, but it can occur at any frequency. The effect continues to be there no matter how low the frequencies of the radiation might be. It is not negligible as some might argue. It is a cumulative effect that requires enough volume of space filled with radiation for this redshifting effect to become detectable and visible.

This is why the Universe is black. Within the distance of the visible universe, the radiation has not yet lost all of its energy. It has redshifted a certain amount to show energy loss, but we can still observe objects. The further you look into the visible universe, the more radiation gets redshifted. Keep looking further back, and eventually, the radiation will have to merge with the rest of the radiation coming in from the Universe. Here, the frequency is well below the visible range in the electromagnetic spectrum, so everything will have to look black right at the edge of the visible universe. Still, even at this range, scientists have detected tiny temperature changes. Does this mean we are seeing the remnants of the Big Bang ready to coalesce into new stars and galaxies? Then again, one can find an opposing interpretation that the changes merely represent clusters of galaxies that are further out beyond our observations and that are helping to heat up some ionised gases surrounding the galaxies.

Seriously, the redshifting effect does not have to be due to the Doppler effect. It could be nothing more than natural energy loss due to radiation colliding with radiation as it travels through space.

Professor Stephen Hawking gave an analogy for a similar energy loss situation. But instead of solid matter emitting radiation, Hawking used solid matter and the emission of gravitational waves to argue his point with readers:

> "General relativity predicts that heavy objects that are moving will cause the emission of gravitational waves, ripples in the curvature of space that travel at the speed of light. These are similar to light waves, which are ripples of the electromagnetic field....Like light, they carry energy away from the objects that emit them. One would therefore expect a system of massive objects to settle down eventually to a stationary state, because the energy in any movement would be carried away by the emission of gravitational waves. (It is rather like dropping a cork into water: at first it bobs up and down a great deal, but as the ripples carry away its energy, it eventually settles down to a stationary state)."

In this quote, if you replace "heavy objects", "gravitational waves", and "cork" with "radiation", the result would be the same energy loss leading to a natural "settling down" of its energy through a redshifting effect.

If this is not enough, looking at the light of exploding stars in distant galaxies is not going to prove an accelerating expansion of the universe. Rather, due to the heightened energy density of the extra radiation coming out of the exploding star, this will act as a more significant solid piece of matter in space to help redshift the radiation more significantly. The redshifting observed in this light need not have to refer to any kind of receding of the galaxy. Come to think of it, how would we know if the source of this radiation moved significantly away from us to create the redshifting effect due to the Doppler effect? You can't. At such distances, any potential redshifting from the Doppler effect is quickly erased through the natural redshifting of the radiation traveling through the heightened mass-energy of a supernova explosion and subsequently through space itself. It might be too difficult to separate out the difference. In this case, how do we know the galaxies are moving away from us? The problem is even worse once we approach the very edge of the visible universe. At this extreme distance, things get really murky. What is out there gets less and less visible and harder to detect. The radiation has redshifted so much, and the energy has spread out enough that it is very hard to detect very distant galaxies and superclusters of galaxies just beyond the visible edge of the universe. Perhaps some of the galaxies could be heading our way and we just don't know it. Obviously, it would be bad for those who believe in an expanding universe. But that's part of the nature of science. We have to challenge our assumptions and make sure that our interpretations are correct. Then, we have to find ways in which to observe the difference in the redshifting due to the Doppler effect and the redshifting due to photon-to-photon collisions in space. Only then can we increase the number of galaxies in our local neighborhood to see which ones are heading in our general direction or are going in another direction. The Andromeda Galaxy is heading our way. An exception to the rule? Probably not. It is likely that many more galaxies will move in our general direction. It is only when we get to the very edge of the visible universe that it becomes virtually impossible to be certain what is happening. Are we living in a finite universe? Or, are we living in an infinite Universe? Your guess will be as good as anyone else's on this very matter.

What does this mean for dark matter and dark energy?

If this unified field idea of light losing energy by redshifting as it travels through space is correct, then in terms of the gravitational field equations, this redshifting of the light must be the thing that is "stretching" the curvature of space-time as Einstein had noted. Makes perfect sense.

Another implication is that all objects sitting on this curvature of space-time need not be expanding (or moving away from us) while light redshifts. In an infinite universe, it is easy to show this is true if one wanted to.

Also, the farther away a galaxy is positioned, the more the radiation emitted by the galaxy has to redshift as it travels through space to reach the eyes of the scientists here on Earth. Exploding stars would merely accelerate the redshifting effect more significantly—but not because the galaxies are racing away from us. This is what we should expect of radiation after making many more collisions and transferring more of its energy to other radiation.

Therefore, the suggestion that the rate of expansion is increasing at greater distances, as if a mysterious dark energy exists in the universe to push everything apart, may actually be an optical illusion according to the Unified Field Theory. Radiation is behaving like the mysterious dark energy repelling all matter, but it isn't repelling anything whatsoever. Radiation from an infinite Universe is still coming in at roughly equal amounts and of the same quality (or frequency) all around the objects and this is what's keeping them roughly where they are other than their usual motions through space.

As for the equally mysterious dark matter talked about by scientists, this is none other than radiation once again behaving in its usual and ordinary solid-matter way, including having its own gravitational field. It is the gravitational field of the radiation that is controlling the speed of rotation of stars around each galaxy such that they move at roughly the same speed.

There is another important implication to consider: how likely is it for any observer positioned anywhere in the universe to prove conclusively through direct observations that the universe is finite or infinite. Indeed, could the universe ever be finite in size at any moment in its long and arduous history if we accept the current scientific view of the Big Bang? Moreover, if we are living in an infinite universe, how can we prove it, too?

Let's discuss this latter point in more detail.

Can the Visible Universe Ever Be Finite?

A perfect vacuum does not allow a Big Bang to occur

There are problems with this inflationary finite universe idea and the Big Bang theory. Apart from not knowing why the universe began at all, we still do not know what was outside this universe just prior to the Big Bang.

Or maybe we do if we apply common sense to the problem.

You see, for anything to explode from a central point, as Lemaître suggested, the universe must have expanded into something much bigger. Already this tells us there had to be a grander Universe for this to happen. So how do we know that the universe is not already the Universe?

The first clue to answering this question is already provided by Einstein's General Theory of Relativity. It concerns the speed of light in space at different densities of the mass-energy permeating all of space. Fortunately, we now know from the Unified Field Theory that this mass-energy is radiation. But let's not worry about this last point for the moment.

As stated in Chapter 2, the speed of light varies in different densities of the mass-energy. Now, this may come as a surprise, but this density issue played an important role just prior to the Big Bang. Not sure? Maybe it is time we paint the picture in our minds in a more obvious way.

Scientists agree that just prior to the Big Bang, the energy and matter of this visible universe (i.e., space and time) had to be tied up inside what most scientists consider to have been a marble-sized ball of, we presume, infinite density at the center, or essentially a *singularity* (just like a black hole, only much smaller and probably more unstable). Already this is telling us something significant. Scientists say that no radiation (i.e., space-time) could have been in the Universe because this black hole had a mass (or gravitational field strength) that would prevent radiation from escaping. Furthermore, any radiation outside of the black hole would be sucked in instantly. Effectively, the Universe outside the black hole had to be devoid of all radiation (and hence space-time through the mass-energy of space). We call this the perfect or true vacuum of space.

Already we should see a problem here. More specifically, both the grander Universe and the visible universe apparently existed as separate entities, with one containing all the energy and mass, and the other completely devoid of radiation. How is this possible? All the mass (or energy) of the universe will not stay in one spot while a perfect vacuum exists all around it. In fact, it is impossible to create a perfect vacuum anywhere in the Universe, not even for an infinitesimal amount of time. If we could entertain ourselves with this impossible scenario, all mass will be torn apart and its energy will travel at the speed of light, which in a perfect vacuum, is, in fact, infinite[13]. If there could be any attempt by some natural means to create this perfect vacuum (even something as powerful as the biggest black hole in the Universe), this vacuum has the infinite power to suddenly draw out the energy in any solid matter or other sources of energy located in any position in the Universe to fill up the perfect vacuum instantly. This includes any mass from solid particles. If a black hole containing the universe could ever exist, it is already in trouble. It cannot keep all the energy and mass of the universe in one spot for any length of time while a perfect vacuum exists all around it, not even for the smallest fraction of a second that you can imagine.

This means everything in our visible universe could not have started from a single point and exploded no matter how small a timeframe we give it. If it ever could, theoretically speaking, have started from a point in space, the mass-energy of the universe should already have covered the entire Universe by now. You only need an infinitesimally small timeframe to cover the infinite Universe given the infinite speed of the mass-energy it will attain in a perfect vacuum.

This brings us to a notable dilemma. How can scientists be sure the universe is, say, 13.7 billion years old? Surely this energy can instantaneously reach an infinite distance in the smallest amount of time imaginable and, with it, create matter. It would be created well beyond the 13.7 billion light years we currently think is the edge of the universe. As for those scientists who say those tiny blobs or points of light in the universal background radiation causing a slight elevation in the temperature is evidence of the primordial gaseous remnants of the Big Bang, energy that has reached an infinite distance means that the blobs could also quite easily represent solid matter. Indeed, this solid matter could already be fully-formed galaxies and stars just sitting out there. So which is it? Primordial gaseous remnants of the Big Bang? Or fully-formed stars and galaxies? Already we have a paradoxical situation developing here.

Even when scientists talk about an incredibly small time frame for our early universe of around a trillionth of a second, this is still an eternity of time for the mass-energy constituting our visible universe to fill an infinite Universe.

---

13   See Chapter 2 for evidence of an experiment to support this speed of light in a perfect vacuum.

Common sense tells us the universe has to be a much bigger place than we can see and must go beyond the faint ripples or blobs in the background radiation. In which case, those ripples or blobs cannot be mere gases ready to coalesce into new galaxies as the scientists claim. There must already be fully formed distant galaxies and stars out there surrounded by the usual ionized gases to help raise the background temperature slightly. That is what these blobs or ripples are all about in reality.

The Universe paradox

Here is another thought experiment. Ask yourself, what would happen if we could magically transport ourselves in an instant using a mathematical "perfect vacuum" wormhole[14] to what we think is the edge of the visible, expanding universe? What will we observe inside these apparent ripples or blobs of empty gases?

Prior to making the trip, we must realize that the light from the edge of the visible universe had to travel, say, 13.7 billion light years of space (filled with other radiation) to reach the Earth. In other words, the light we scrutinize on Earth is already 13.7 billion years old. Remember, it has taken that long because light has to travel through the natural density of radiation in space. Okay. So let us imagine we could instantly transport ourselves to one of these blobs at the edge of the visible universe using our hypothetical wormhole. We would arrive at the blob 13.7 billion years into its future. This means we can no longer assume the blob is a mixture of gases ready to coalesce into a galaxy; it will by now be a fully formed and probably ancient galaxy (or galaxies) like our own.

Already we have a paradox. On Earth, scientists claim the blobs at the edge of the universe are comprised of primordial gases from the beginning of the Big Bang ready to come together to form the next generation of galaxies and stars. However, when we arrive instantaneously, we discover the blobs do contain galaxies and stars. Something is clearly not right. There is no way the edge of the universe can be 13.7 billion light years away from Earth. It has to be much farther than this.

---

14   This is known as the *Einstein-Rosen bridge*, a hypothetical "perfect vacuum" tube for allowing any mass-energy to accelerate to infinite speeds.

In fact, the universe should be at least 13.7 x 2 (or 27.4) billion years old. Why? Because this is how much extra time the radiation and the matter from the Big Bang has had to reach what we perceive to be the edge of the universe from this new vantage point (assuming radiation only moves at 300,000 km/s).

So, we look out again and discover yet more of these ripples (or blobs of light) at the edge of the universe after another 13.7 billion years of extra time for light and, hence, matter to reach there. But hold on! Are we not falling into the same trap again? Surely those blobs are more galaxies and stars. If we are not at the center of the universe, then this is probably more of a reason to suspect that these blobs are more galaxies and stars.

As this thought experiment reveals, we really have absolutely no idea where the true boundary of the universe lies (let alone that of the grander Universe). Any estimate on the age of the universe, let alone the grander Universe, even if we could travel to the edge of the universe and make the observations, has to be considered a complete guess.

We may think that, on Earth, the edge of the universe is 13.7 billion light years away based on what our eyes and instruments can see and, therefore, that we live in a universe that is only 13.7 billion years old based on how long it has taken light to reach us; in reality, however, we already know it has to be a lot more than 13.7 billion light years after traveling what we think is half the diameter of the observable universe to reach where we think the edge is (assuming we are at or very near the center of the universe). If we travel again, and again, to reach where we think the edge is, can we be sure the universe is not 60, 80, or 100 billion years old, or even older?

Get the picture?

In other words, it is pointless to ask how old the universe is if we are clueless about where the edge of the universe actually is. The visible universe is throwing at us a paradox that we cannot resolve to a single answer.

Is our universe a bubble among many other bubbles?

There are many universes sprouting in the Universe

Understandably, some scientists might feel a little uncomfortable with these arguments. So, to get around the problem, it has been recently proposed that this grand singular Universe may consist of many universes such as our own sprouting out from a giant space-time "foam" just like the bubbles in a bubble bath. At least this avoids our visible universe being at the center of the Universe and helps to explain why matter has not suddenly disappeared to infinite distances.

Michio Kaku, a theoretical physicist at City College of New York and proponent of this multiple "expanding" universe idea said:

> "The multiverse is like a bubble bath [where each bubble represents a universe. There are] multiple universes bubbling, colliding and budding off each other [all the time]."

Very creative—until we apply some rational thinking to the idea and keep in mind the known behaviors of radiation in space at different energy densities.

For example, how do we know other universes have not instantaneously merged into ours to form the Universe? If our visible universe is meant to be so brand new, what is stopping light and matter from self-accelerating to infinite speeds into the unknown true vacuum outside and merging into these other universes we are currently unable to see (and eventually decelerating in the new mass-energy of space in these other universes)? In fact, what we should be seeing in the radiation is not an acceleration of the galaxies receding from us. We should see a slowing down—if not, we should already have reached a steady state with all the other universes.

Such a proposal is interesting, but it does not necessarily prove without a doubt whether the visible universe we live in is finite or infinite or how old it is. Nor does it provide an alternative explanation for the observations of a redshifting effect in the light from distant galaxies (just to be sure that our interpretation of the observations is indeed correct).

Even more concerning is the disconcerting "center of the universe" issue in terms of time itself to contend with, according to the work of Cox and Loeb.

What is going to happen to our visible universe?

Researchers T. J. Cox and Abraham Loeb said that they think the only thing to watch in this universe 5.1 billion years from now will be our Milkomeda—the new name given for the galaxy formed after the merger of the Andromeda Galaxy with the Milky Way—and a few local groups of stars. In other words, scientists today believe our visible universe of galaxies will disappear from sight into the unknown and unseen grander Universe and eventually be pitch-black except for our new galaxy and a few local groups of stars caught in the gravitational field of our new galaxy. Give it more time, and eventually even our galaxy will fall apart, stars will stop forming, rocks and ice will evaporate as the energy in the subatomic particles slow down and emit their energy. All life in the Universe will cease to exist and we will freeze ourselves out of existence into pure electromagnetic energy and fade away.

Why?

This suggestion follows from the work done by professors Brian Schmidt, Adam Reiss, and Saul Perlmutter as discussed earlier, which involved relying on a mysterious dark energy—whatever it may be in reality—for explaining the supposed continual expansion of the universe.

We are now getting into some really bizarre implications about our universe, and all because we have accepted Hubble's assumption that everything in the universe is expanding.

But why should this moment in time give us the remarkable opportunity to see galaxies through our telescopes as we speak, when only 13.7 billion years ago everything was right up against our noses, and in the next 5.1 billion years from now, we will not? Are scientists trying to put us at the center of the universe again, this time in terms of time itself?

Furthermore, the Unified Field Theory suggests that the dark energy is an optical illusion. This so-called expansion is nothing more than natural energy losses by radiation as it settles back down to the normal ambient conditions in the middle of space after constant collisions with other radiation.

Such proposals and interpretations of the evidence by scientists for a finite and expanding universe are interesting but do not necessarily get us closer to the truth about our universe we live in (let alone the grander Universe). Scientists could be interpreting the evidence in the wrong way to support a theory that may end up being completely wrong.

The Big Bang was not a sudden expansion

So now, some scientists have gone one step further by suggesting the Big Bang was probably not a sudden expansion like an explosion but rather a gentle expansion, which helps to avoid the idea of matter and energy racing away into the infinite Universe at infinite speeds. It is the only way to explain why we can now see galaxies near our Milky Way and those mysterious blobs at the edge of the visible universe with our telescopes. Unfortunately for the scientists, a perfect vacuum in the Universe will not allow for a gentle expansion of matter and energy. It will be a violent and instantaneous expansion. So technically it is more correct to say it would be an explosion. And any gentle expansion must be evidence of the visible universe already merging with other "universes" or the grander Universe already contains mass-energy. And over time, that expansion should actually be slowing down, not accelerating.

The misconceptions continue unabated.

Better still, why don't the scientists just start from scratch and look at the original observations of the redshifting effect in the light once again to make sure our fundamental interpretation is truly correct? It is the only way to be sure about this.

Can We Truly Give an Age and Size for the Paradoxical Universe?

In that case, what would Hubble have stated about the visible universe if he had the hindsight of the Unified Field Theory and the way the ordinary matter we call radiation behaves in space? Would he still think the universe is finite? Probably not. It is likely Hubble would have stated that the galaxies are still moving at their usual speeds in any direction no matter how far away they are from the Earth. Therefore, the universe could potentially be infinite for all intents and purposes. Of course proving this infinite idea would still be as difficult as proving the universe to be finite. Why? Because the cosmologists' attempts to determine the actual speed of the galaxies using the Doppler theory may be futile. You see, light from these distant galaxies is already losing energy as it travels through space. How would one know how much of this redshifting effect is due to the Doppler effect at the point the radiation was emitted by a distant object? Furthermore, if the universe is much larger than we have anticipated and there is a possibility of it could be slowing its expansion, it may take a massive amount of time to wait until we can take another measurement (minus the Doppler effect if we can measure it) and see if the expansion is slowing down. If the Universe is extremely large and very old, it is potentially possible that all the time we have left on Earth may not be enough to reveal the difference.

The only way one can be certain is to travel to those galaxies near the edge of the visible universe to see what is really happening. In a physical sense, that is an impossibility. We will have to hope that there are advanced alien civilizations throughout the universe that have had enough time to pass on information to each other to help everyone gather a more complete picture.

Until we can become part of the universal network of intelligent scientists out there, is it true to say the universe is expanding, contracting, belching or whatever it is doing? Given what we have learned from the Unified Field Theory and what light is doing in space, the answer must be that we don't know exactly what the universe is doing. We can only guess.

As for the age and size of the universe (the part that we can observe), the only thing we can say about this universe with any real confidence is that it is a truly big and very old place. Exactly how big and how old (are we talking about the universe, or the much grander Universe?) and whether it is expanding or contracting, finite or infinite, is anyone's guess until we gather more evidence and *interpret* that evidence correctly.

CHAPTER 7

The Secret to Immortality

Anyone who has never made a mistake has never tried anything new.

—Albert Einstein[1]

IGHT, or electromagnetic radiation, is the most pervasive substance known to science. Not only does this energy seem to reach out to all matter in the Universe[2] in a kind of invisible electromagnetic glue or "guiding hand" binding, influencing, or helping all matter to follow certain physical laws of the Universe, but it can also penetrate matter to its deepest level. Nothing can escape its reach—not even the living cells in our body.

This naturally raises the question of whether Einstein's Unified Field Theory has an influence on the aging process. It seems appropriate that we look at this more closely.

---

1   The general consensus is that this quote is attributed to Einstein.
2   We will assume we live in an infinite Universe and not a finite universe when capitalizing this word.

Defining the Term "Aging"

According to the 1992 edition of *The Encyclopedia Britannica,* "aging" is defined as:

> "The progressive changes that take place in a cell, a tissue, an organ system, a total organism, or a group of organisms with the passage of time."

In Wikipedia in 2010, the definition was:

> **"Ageing** (British and Australian English) or **aging** (American and Canadian English) is the accumulation of changes in an organism or object over time. Aging in humans refers to a multidimensional process of physical, psychological, and social change."[3]

The aging process in action, showing the hands of an 86-year-old woman.

---

3   Wikipedia definition available from http://en.wikipedia.org/wiki/Aging_process.

At the BioMed Central Ltd website, the definition is as follows:

"Aging is the accumulation of damage to somatic cells, leading to cellular dysfunction, and culminates in organ dysfunction and an increased vulnerability to death."[4]

Whatever causes all living things to age, these definitions share a common theme: the idea of "change", and how this change accumulates in the body over time.

We Live in Constant Change

Albert Einstein was acutely aware of this change, not just in terms of his own age, but throughout the Universe[5]. He said, in a somewhat complicated and scientific way, that there is no such thing as an absolute frame of reference with respect to all things in the Universe. For example, you might be happily at rest while reading this book (termed a "reference frame") and later look up at the world and see constant change, while another observer sitting in space above the Earth's atmosphere in his own reference frame will look at you and think you are in a state of constant change. Why? It is because the Earth rotates and you are moving with it. You are being impacted by various forms of matter, including radiation. If you think you are not moving in your reference frame, think again. You are moving and in a state of constant change.

As oxymoronic as this may sound, it remains a fundamental and unchanging truth. The one constant or unchanging idea about this Universe is the idea of change itself. Therefore, as long as you are constantly moving with respect to other things, there will be plenty of moments when you can expect to interfere with or be interfered by certain things in the environment, resulting in forces being exerted on you and your living cells, forcing you to change during the interference (or collision).

---

4 Izaks & Westendorp 2003 (licensee BioMed Central Ltd). Available online at http://www.biomedcentral.com/1471-2318/3/7.

5 Given what we learned in Chapter 6, we will assume the universe (the visible part) is much larger than we can observe with our eyes and instruments. In which case, the totality of everything in the visible universe and in the invisible universe (i.e., the part that is beyond the visible universe) will be called the Universe.

It is this interference of your living cells by whatever is within the body and/or outside of it, leading to some kind of change, that is the cause of the aging process.

Why Do We Change?

Anything that is solid and physical in the Universe must undergo changes. How much we change depends on whether you are a hydrogen atom floating around through the Universe, or the kind of complex human being you are today. Generally, the more complex we become, the more likely we will use technology, the food we eat, and anything else we can find in the environment to reduce the amount of change we experience. No matter what anyone does, there will always be changes.

Change is inevitable. Things get pushed around and eventually interfere with other things. Some of the interference can be beneficial in the sense that you have become more intelligent and reached the complex creature that you are today. This is why, in evolutionary terms, our bodies have advanced from simple little atoms floating in space more than 6 billion years ago to the single-celled organisms we were nearly 4.3 billion years ago, to the two-legged, big-brained creatures we are now. The forces we were subjected to over millions of years by our environment and within our bodies have shaped our bodies and brain, and ultimately our thinking, to what we are today. Thanks to all these changes we had to endure, adapt, and eventually find a solution to handle the changes, we can finally question things and ponder our purpose in this Universe, as well as the meaning and aim of this Universe.

Still, you might be thinking, "I'm happy with who I am. Why do I need to change?"

Whether you like it or not, you must experience change while you are alive and kicking in this Universe. You can whinge to your heart's content saying things like, "I won't learn anymore and will keep myself cocooned in this room!" But the Universe has an uncanny habit of finding a way to interfere with what you are doing and continuing to push you to change. The Universe will not stop simply because you choose either not to learn or to keep yourself protected in some location.

On the other hand, if the question is more about why change must continue into the foreseeable future, which would see your living cells and the rest of your body age over time until you face the moment of your mortality as well as create evolution, then we enter a more metaphysical realm. This would require the extensive right-brain (i.e., visual and creative) skills of those people involved in the religious and artistic worlds for an answer.

In case we are not aware of it, the right side of the brain helps certain people to pick up hidden, large-scale patterns in groups of observable, small-scale patterns or events. Religious leaders who meditate on the Universe and life in general and the more creative individuals are particularly sensitive to these large-scale and invisible patterns. If all these patterns could be summarized in a single statement, it would be that change in the future is necessary to make us learn a great thing so we may approach a more stable and "balanced" state. To the religious person, the ultimate state to aim for may be called "God". In the event that we should somehow achieve this state in the ultimate sense, we would fully understand who or what God is, discover our ultimate purpose in this Universe, why the Universe exists, how it came to exist, where we and the Universe came from, what we and the Universe are doing today, and where we and the Universe are going in the future at any time and place, as well as how to survive indefinitely in this Universe (i.e., true immortality). Such an ultimate state would effectively make us God. In other words, we would be just like what we see in the following quote from 1 Timothy 6:16 in the Bible for religious people:

> "[God] who alone possesses immortality and dwells in unapproachable light; whom no man has seen or can see."

Until we reach this state, however, we have to learn as part of the process of adapting to the change and understanding this Universe—all of which takes time. Time for which we do not have a lot of, as revealed by the fact that our body and brain are not able to live forever and there is much to learn about how to control this aging process before we can potentially find a way to live for an extremely long time. While we don't know all the factors affecting the aging process, we all have to face the issue of death.

But why do we have death at all? Wouldn't this be an absolute fizzer in helping us get to this ultimate destination? Perhaps. But another thing that creative and spiritual people have noticed is a continuous cycling and recycling of matter and life. You might die tomorrow, but something about this Universe informs these people that we will come back to start again. How exactly this is achieved is not yet understood. Further work by the R-brain people is necessary to look deeper into this problem. The only thing we can do is ensure the knowledge remains intact and is accessible and available in the next life by some means. Then, hopefully, the journey of experiencing further change will be more controlled. If we learn enough, we may enter a greater period of stability where living a very long time is a reality. Indeed, the more you experience greater and longer-lasting stability and happiness because there is less change from the environment (and, hence, less chance of suffering if the change is not favorable) thanks to your knowledge and physical abilities to handle the change, the closer to this ultimate goal you will come. Then, the experience of death is delayed, which means you will live longer.

Death only exists because we are not perfect. Until we have learned everything there is to know about the Universe and ways to handle the changes it throws at us, we will always be children in the great classroom of the Universe. Until all knowledge has been acquired and understood and implemented in ways to deal with the natural forces that make us change (whether we like it or not), we will always experience death.

Thus, the only way to extend our lives (or delay the experience of death) is to learn all the things that forces us to change, find ways to adapt and handle the change, and to repeat this cycle of change and stability enough times in this life and in every other life you will have after death for our body and genetic code to adapt and improve, and for our brain to be large enough to gather enough knowledge and experiences so we can know how to handle the change and so become stable and ageless.

Does this mean that, at the end of evolution (if such a situation can occur in reality), that you will become the ultimate stable and happy living immortal holding the answers to everything, including why you are here, after knowing you can adapt to any environment the Universe can throw at you with no further need for the experience of constant change?

Probably.

But in this case would you know you are God?

No. You really do not know for sure if you have reached this ultimate "balanced" state. Another mystery of the Universe is that it does not allow any living organism to know if it is God. Given that we are all constantly striving to reach this state, it will not matter how advanced our technology or knowledge is, or how long we have lived in this Universe, there is only one God. The Universe ensures that there can be only one God. This one and only God is something we have to approach, and that means we must continually learn all we can about this Universe. If you should find any limitations in your knowledge and/or physical abilities that are preventing you from knowing how to handle a particular change or solve any kind of problem you have or can imagine, then you haven't quite reached the ultimate state of being God. Until you know (and are acknowledged by all living things that see you), it is vital that you adhere to the fundamental law of the Universe, which is that there can be only one God. The rest is up to you to discover.

Although you have not learned everything there is to know to become a true God, it is better that you do not claim to be God. To make sure of this, you will not know for sure at what point you have reached this ultimate level, which is probably how the Universe is designed to ensure there is only one God. The only clue you have in knowing you are not God is the fact that you will die eventually. Therefore, you must always be humble in your limitations and see yourself as a curious child in the Universe ready to learn new and astonishing things, and to believe there is something else beyond death. Let the Universe be your teacher. And never assume there is nothing more you can learn.

Yet some readers still might ask, "What's the point of changing and learning from the changes if we can't or don't know when we have reached this ultimate state of stability to be called God?"

Nobody knows the answer to this great question. But if one could suggest a possible clue, it may be that we do not need to become God. Without breaking the fundamental law of the Universe, there is absolutely nothing stopping us from approaching this ultimate state of stability, and realizing this is probably more than adequate for all living things to be happy and experience enough of this Universe for a long time. Afterwards, the great passage of time could be enough to see us wanting to shed our biological clothes we are wearing and try on new ones in our next life in order to have a closer and more intimate experience of God.

The only thing we can say on this issue is that there is nothing stopping you from getting closer to God through the changes you make now and achieve greater stability. As you approach this goal, the knowledge you will gather will help you to achieve a greater expansion of your lifespan. Something that we hope you will share the secrets with others. With your youthful and incredibly long life, you will have plenty of time to experience and learn so many more things about this Universe and yet still be content even when you are ready to die and move on.

But if we are still not content with life and are not ready to die and move on, then maybe the Universe is trying to teach us something else. It is possible that we all need to be mortal in order for other life to emerge into the Universe and be given the opportunity to experience and learn from their experiences when living in this mysterious Universe just as we have. The only thing that may benefit other life is the recording of your knowledge so that they may approach more quickly the thing we are all aiming for and, therefore, experience greater stability and joy for even longer. Then, at some point in the future, it should be your turn to re-experience once again what it is like living in this Universe—except the next time (if everyone has followed the principle of love and retained the knowledge from previous generations) you will benefit from greater stability and the knowledge of a longer life from those who lived before you. So long as everyone was loved when they were alive, encouraged to achieve great things, and share that knowledge with everyone, this should happen every time we come back into a new life.

Of course, for the more rational types, no doubt they will consider this mere speculation. A completely understandable position considering they cannot see with their eyes what happens after death, let alone all the hidden patterns picked up by the more creative and religious types.

One thing is certain: The necessity to change seems to be an inherent part of the physical and real Universe in which we live. We have no choice. The living cells in our bodies and in ourselves have no choice but to change. Therefore, the questions to ask are which changes are likely to be detrimental to how we look and feel over time as we age, and how do we control these changes?

What Causes the Cells to Change?

Scientists have identified a number of factors that seem to affect or control the rate of change in the cells of a living organism.

Positive emotions and thinking

One such factor that influences the rate of change in living cells is our general state of psychological and emotional well-being. For example, a positive attitude toward life and regularly thinking we are young as well as maintaining an active brain to solve problems are thought to prolong the aging process, perhaps by secreting the right sorts of useful hormones in the body.

"The brain thrives on stimulation", says Dr. Marian Diamond of the University of California, Berkeley, "and it dies without it. Use it or lose it."[6]

Furthermore, if the brain finds the stimulation positive and wants to learn more, the positive attitude toward learning has an impact on the way our bodies age over time.

In an experiment conducted in 1979, a group of male volunteers aged over 75 years, who were living under the care of someone else, were asked to live independently in homes set up in a way to remind the volunteers of their pasts. The homes contained 1950s-designed furniture, a black-and-white TVs, record players, fridges, and anything else the volunteers remembered were positive experiences and helped them to relax.

---

6   Batten 1984, p.40.

The aim was to get the men to think and behave like they were living during the 1950s at a time when they were much younger and most likely happier.

Professor Ellen Langer of Harvard University, the person who headed the experiment to test the mind over the body as a possible way of controlling the aging process, said:

"These people [the men] looked like they were on their last legs. So much so that I said to my students, 'Why are we doing this? This is too risky. I was going to take over their lives basically for a week'

...We created this environment they were going to be totally immersed in. It was a timeless retreat that we had transformed. And so for a full week they would be living there as if it was that earlier time.

...As soon as we got off the bus, I told them that they were in charge of their suitcases, getting them up to their rooms, they could move them an inch at a time, they could unpack them right at the bus and take up a shirt at a time. Just think about the difference in how these people were treated: by me, with the assumption that they could do everything; versus treated like a little kid.

There was nobody bathing them. They were in all ways taking care of themselves as they would have and did, say, 20 years earlier."[7]

The experiment may have lasted a week, but Professor Langer noticed significant improvements in health and observed a range of autonomous activities that would no longer be described as symptoms of the aging process. The professor stated:

"We got a difference in their dexterity; a difference in their joint flexibility; they were able to move faster; they stood taller; their cognitive abilities improved; [and] their blood pressures dropped.

Indeed, an IQ test on the men revealed an astounding 63 percent improvement compared to the same test performed just prior to the experiment. Add to this the improvement in

7   Quote taken from the BBC documentary *Don't Grow Old* (2009).

the men's eyesight and hearing abilities, and how all the men had learned to do things on their own without assistance from anyone else, and one would have to conclude the symptoms of aging had been reversed."

The experiment is remarkable, for it suggests that positive emotions and thinking can reverse many, if not all, of the symptoms people tend to associate with the aging process. As Professor Langer said:

"These findings are quite astounding. Remember, old people are only suppose to get worse. Most people don't assume vision will improve, hearing will improve. Certainly not cognitive abilities. There are certainly symptoms of arthritis diminishing. And all of this from them [the men] just living as if they were younger for a week's time."

In conclusion, it seems how healthy you are and eventually how old you feel are determined by how you think and the positive emotions you can create. As Professor Langer said:

"Your views of your own aging are going to largely determine how you age. If you view yourself as somebody who is going to fall apart, you will fall apart. You will probably live just as long as you think that you are suppose to live. That again we have enormous control over our health and well-being that we are only beginning to become aware of."

In some sense, it is almost as if the experiment is suggesting that society should abandon the idea of having birthdays every year or other important milestones in our lives so we are not made acutely aware of our ages. Instead, we should learn to think positively about what we are doing, find time to do things on our own, and believe we can live as long as we want to participate in life. Then the body and mind will follow.

Speaking of hormones for longevity, in 1969, Dr. W. Donner Denckla, an endocrinologist formerly with the U.S. National Institutes of Health (NIH) at Harvard University, noticed how, as we grow older, the pituitary gland releases more of a "decreasing-oxygen-consumption hormone" (DECO) that inhibits the ability of cells to use thyroxine—a chemical needed to control the rate at which cells convert food to energy. This could reduce our lifespan. To test this theory, Denckla removed the pituitary gland from rats and measured their lifespan among other observations. What Denckla discovered is that the loss of this hormone is said to increase the youthful functions of rats and their lifespans.

Does this mean the hormone is solely responsible for the aging process? Or could it be the lowering of oxygen in the body that is the deciding factor in reducing the lifespan? Or is it the brain that controls aging? As Denckla said:

> "Look at it this way. The carpenter is the man who has the hammer in his hand to drive the nail. The pituitary [gland] is the hammer, but it sure isn't the carpenter. Where the carpenter is, I wish I knew. It's probably in the brain."[8]

As another example, the hormone-producing endocrine system is linked to the immune system, which assists in combating diseases such as cancer. Of considerable interest to scientists is the thymus gland. Dr. Allan Goldstein, chairman of the department of biochemistry at the George Washington University School of Medicine and Health Sciences, says the thymus gland secretes hormones he calls "thymosins", which appear to restore the immune system in old animals as needed to handle viruses and bacteria entering the body.

Goldstein says thymosins control the production and function of three different types of white blood cells, or lymphocytes, known as T cells. One type of T cells—the "T-killer cells"—is known to produce lymphokines. Lymphokines are the immune system's own natural drugs for fighting diseases such as cancer (i.e., they destroy virus-infected cells as well as cancerous ones). The second type of T cell is called "T-helper

---

8   Batten 1984, p.40.

cells". Their purpose is to produce antibodies. The third type, "T-suppressor cells", helps to regulate the immune system and prevent it from attacking the body's own living tissues.

Goldstein and his research team also discovered that thymosins stimulate hormones in the brain, including the "feel good" hormone normally released during relaxation or sex called beta endorphin.

Over time, however, the thymus gland shrinks, the production of thymosins goes down, and the ability of the immune system to combat diseases is greatly reduced. Studies by Dr. William Ershler of the University of Vermont School of Medicine and others on the effects of injecting thymosins into the elderly are most promising. In trials using the hormone, improvements in various vaccine and cancer treatments were observed when white blood cells in older patients were given a dosage of thymosins.

"In the future," Goldstein says, "I think that thymosins may play a very important role in preventive medicine."[9]

Thymosins are not the only hormone to go down in quantity over time. Today, other doctors have realized that a variety of hormones[10] do fall in quantity quite dramatically as a living creature ages over time. On further analysis, scientists have discovered a way to restore the levels of these hormones by introducing another hormone designed to make living cells divide and grow. In humans, it is called the human growth hormone (HGH).

The idea is that if humans could inject themselves with some HGH as part of a controlled hormone-replacement program, youthful characteristics such as greater strength in the muscles and more elastic skin should become apparent within a few months. Well, that's the theory anyway. In reality, most individuals undergoing the hormone-replacement program have noticed improved muscle strength and better skin tone. Some have claimed to be more alert and active under this program. However, for others, this is not so clear, suggesting the success of this program depends on the quantity[11] of HGH injected into the body, and how well the individuals look after their health (e.g.,

---

9   Batten 1984, p.42.
10   Including hormones that help the body to remove waste left behind by cells.
11   High doses of hormones can reduce the safety and benefits to the body. And choosing certain synthetic hormones from pharmaceutical companies instead of the bio-identical hormones from plant-based materials can increase the risk of cancer (e.g., prostate and breast) because of the way the chemical structure of these synthetics types is changed slightly to make them patentable.

do they stay out in the sun most of the time?) and eat the right diet in the early parts of their lives to gain the most benefit from HGH in later life. In other words, the cells have to be in excellent condition and healthy enough for this hormone replacement program to be useful.

In fact, in other studies, scientists noticed how the prolonged use of growth hormones can increase the likelihood of cancer developing in the body. Why? It essentially comes down to the quality of original cells for dividing and growing. If these original cells are not in excellent condition in the first place, especially at the genetic level, further cell division and growth would only increase the risk of further damage taking place. And should the damage be too great, the cells may become cancerous in nature (i.e., they grow out of control and at a rapid pace).

Mitochondria regeneration

Whether it is the cause of aging or a natural symptom of the aging process, another aspect of the body demonstrated to change over time is the amount of energy generated by special biochemical powerhouses known as *mitochondria*. Found in every living cell, mitochondria provide all the essential energy required in order to run the cells and facilitate their work. Should the energy levels drop as a result of mitochondrial dysfunction, as is usually the case when we age, certain diseases leading to heart failure, neurodegeneration, and diabetes tend to develop. To be more specific, ineffective mitochondrial activity in living cells can lead to:[12]

- Neural cognitive dysfunction;
- Inflammation of vascular systems leading to hypertension, heart attack and stroke;
- Increased fat storage in the liver;
- Increased visceral fat storage, i.e., belly fat;
- Increased blood sugar levels, insulin resistance, and metabolic syndrome; and
- Increased fatigue and loss of muscle strength.

---

12   http://alivebynature.com/about-niagen/

To determine whether this could be an important factor in the aging process, scientists from the University of NSW, Sydney, headed by biologist Professor David Sinclair, focused on the chemical pathways leading to the formation of energy within the mitochondria. Of particular interest is a chemical known as nicotinamide adenine dinucleotide (NAD). Considered to be an essential molecule in energy metabolism, including communications between the cell nuclei and the mitochondria among other roles, NAD has been measured to decrease in significant levels within the body over time. When NAD drops, it affects the activity of a group of important enzymes referred to as sirtuins, including SIRT1, which is considered to be a beneficial chemical in the formation of new mitochondria, and SIRT3, which is known to keep mitochondria running smoothly. When all sirtuins work together, they act as master regulators to help divert the energy into either the cellular division and growth phase or cellular preservation in times of famine (i.e., a restriction in the number of calories in our foods), which causes the body to find ways to preserve itself and protect against diseases. As Sinclair said:

> "They [sirtuins] are the body's natural defense against disease."[13]

At any rate, a reduction in NAD has a notable effect on these sirtuins to the point where mitochondria start to deteriorate and ultimately cause a cell to lose its zest. Whether NAD is the fundamental cause for mitochondrial deterioration or merely a consequence of some other factor that affects NAD, the scientists have conducted an experiment to determine what happens when the levels of NAD are restored.

The work into NAD began with the release of a high-profile paper titled, "Declining NAD(+) induces a Pseudohypoxic State Disrupting Nuclear-mitochondrial Communication during Aging", published on December 19, 2013, in the *PubMed* journal. The authors, Sinclair et al., discussed the results of an experiment involving elderly male mice that were the equivalent of a human of 60 years of age and what happened when NAD levels were restored. A chemical was added to the drinking water called nicotimamide mononucleotide (NMN)—naturally found in

---

13   Zubrzycki 2015. Article available at http://www.smh.com.au/good-weekend/never-say-die-david-sinclairs-antiageing-quest-20150916-gjocnm.html as of December 2016.

cells and known to boost the levels of NAD. Combined with a high-fat diet to mimic symptoms of old age, including diabetes and obesity, mice that received doses of NMN demonstrated, in a matter of just one week, measurable observations of youthful behavior (the equivalent of 20-year-old humans), including more energy to perform exercises and an average of 60 percent reduction in the weight. More time was still needed to see if muscle strength would improve, but it seemed natural and expected that this would occur should the mice continue exercising. Even after repeating the experiment, the words used by the scientists described the NMN's effects on energy metabolism was "nothing short of astonishing".

Professor David Sinclair of the University of NSW.

The mice used in the experiment were male. In another experiment, female mice were used. The results again followed those of the male counterparts but with one added advantage for the females. As Sinclair reported on the ABC Catalyst science program on June 21, 2016:

> "So most of our experiments were done on male mice. They just happen to be simpler biochemically than females. Then we started to study female mice and what we noticed was that the old mice that were given our molecule [i.e., NMN] were able to reproduce for longer."

Previously, Sinclair worked on a substance found in red wine called "resveratrol" where he discovered its positive effects on one sirtuin enzyme. Later, he saw how NAD works on the entire group of situins with even greater benefits. As a result, the work encouraged Sinclair and his colleagues to find ways to boost NAD levels in humans to see what would happen. Sinclair said:

> "Exercising and dieting can boost NAD levels, but as we age, our body makes less and less NAD. Resveratrol works on just one [situin enzyme], whereas NAD works on all seven. NAD is the fuel for these enzymes. Think of it as the petrol and resveratrol as the accelerator pedal. You need the petrol but you also need the accelerator, so if you have both it's even better."[14]

NMN is not available to humans as a consumer product. However, a precursor to NAD, known as Nicotinamide Riboside (NR), is found in trace amounts in milk. Its presence is known to boost NAD levels in the same way as NMN. Now a half-dozen Nobel laureates and distinguished scientists are taking on a "never say die" attitude by working closely with two pharmaceutical companies—ChromaDex and Elysium Health—to manufacture NR supplements to see if NAD can be returned to normal healthy levels for humans. Mayo Clinic's Jim Kirkland, a respected leader in geroscience; Lee Hood, a biotech pioneer; and Roger Kornberg, a Stanford professor who won the Chemistry prize in 2006, are among the big scientific names to get involved in this work.

In October 2016, ChromaDex announced it had conducted a clinical trial of NR supplements on humans using its own product called "Niagen". Led by PhD. Professor Charles Brenner and Roy J. Carver, Chair of Biochemistry at the University of Iowa Carver College of Medicine, in collaboration with colleagues at Queens University Belfast and ChromaDex Corp., 12 healthy human subjects (6 men and 6 women) were each given a single oral dose of 100mg, 300mg, or 1,000mg of NR in separate trials with a gap of a week between each trial to allow the body to return NAD to normal levels. Analysis of blood and urine samples of the subjects in Brenner's labs revealed that

---

14   Zubrzycki 2015.

the NAD metabolism increased by amounts directly related to the dose. More importantly, the higher dosage rates showed no serious side effects. Today, the company sees NR as so beneficial that it has described it as a vitamin. More specifically, it has been given the name *vitamin B3*.

As for Elysium, it is hedging its bet that a synergistic combination of NR, along with another natural substance, i.e., pterostilbene, which is found in blueberries and grapes, will improve absorption of NR leading to a boost in NAD levels in humans through its own supplements. Also, recent test tube and rodent studies into pterostilbene has shown evidence of improving brain function and preventing cancer and heart disease as an added benefit. The NR supplement it produces is called BASIS.

In February 2015, Elysium conducted what is probably the earliest human trial of NR using BASIS. Each capsule contained 250mg of the Chromadex Niagen brand of NR combined with 50mg of Chromadex Pterostilbene. Despite early testing of this product, the results were immediately apparent. The company claims "a single dose of NR resulted in statistically significant increases [in NAD levels for humans]". Presumably this would have translated into improved energy levels for the subjects (although any weight reduction, suggestions of improved memory, and other benefits would require extra time to observe and measure). Nonetheless, when the experiment continued for 8 weeks, all 120 elderly human subjects who participated in the study had maintained a 40 percent increase in NAD levels over the period of the trial. In another trial lasting 4 weeks, the daily dose of 500mg Niagen had resulted in a 90 percent increase in NAD levels for the subjects throughout the period of the trial. Any higher doses and measurements showed that the body was able to set a ceiling on the level of NAD with the excess NR secreted from the body through urine. The company has completed its study and published[15] its findings on November 24, 2017, in the *Nature Partner Journal of Aging and Mechanisms of Disease*.

Both the Chromadex and Elysium studies show that between 250 and 500 mg of NR provides the optimum NAD levels for the human body.

---

15   Dellinger et al 2017, pp.1-9. A summary available at https://www.elysiumhealth.com/en-us/basis/human-clinical-trial-results.

Further work is continuing into this interesting field of study to determine just how successful human trials of NR supplements will be in reversing mitochondrial deterioration. In particular, whether increases in NAD results in reversing age-related diseases such as diabetes and heart disease. For this, a much longer human trial will be necessary.

Free radicals

Researchers believe another factor controlling the aging process concerns the way our bodies are damaged by charged molecular waste products called "free radicals".

Dr. Roy Walford of the UCLA Medical School has studied these biochemical misfits, describing free radicals as "the great white sharks of the biochemical seas". Free radicals are unstable, mostly negatively charged molecules. They are formed from the by-products of metabolism. In other words, for your body to perform a task such as moving a finger or creating a thought, oxygen in the atmosphere has to be absorbed and used within biochemical reactions inside the living cells to create energy. Once energy is produced, the task—such as contracting muscles or firing nerve impulses in the brain—can be performed. As a result of creating this energy, a certain amount of free radicals are formed.

As Professor Arlan Richardson of the University of Texas explains it:

> "One of the important components of all living organisms is the need for oxygen. And the reason we need oxygen is that we use oxygen to essentially generate energy so that we can live. And so when we take that oxygen in and we burn it, we have one of the by-products that comes from this. It is not only do we get the energy, but we also have radicals or oxidative stress.
>
> All of life actually is a kind of a trade off. You know, to get this energy, you essentially have to pay a price. And the price from the oxidative stress theory of aging, the price would be that you're eventually going to age."[16]

---

16   Quote from the BBC documentary *Don't Grow Old* (2009).

Free radicals perform their infamous oxidative damage to cell membranes and DNA in a manner not unlike the way oxygen in the air oxidizes iron and turns it into rust. Walford believes the highly persistent nature of free radicals throughout life makes it rather difficult for the body to repair all the damage properly, and, over time, ravaged cells accumulate within our bodies in both a genetic and non-genetic wear-and-tear process, eventually producing the effects of aging.

The body's best defense against these marauding chemical invaders, Walford says, lies in enzymes that break down or scavenge them. For example, scientists have already confirmed that certain substances consumed from plant foods can assist these enzymes in reducing the number of free radicals in the body. Vitamin C is one example of such a substance. Scientists have also confirmed that modifying portions of the genetic code in DNA can regulate some of these helpful scavengers.

Another way to reduce free radicals is to drink plenty of fresh and clean water to help neutralize the electric charge carried by these free radicals.

Support for the idea that a vitamin-rich diet low in calories can potentially reduce the symptoms of aging by regularly mopping up free radicals can be seen in those individuals who have made it a life-long plan to consume vitamin-rich, mostly plant-based, foods. The diet involves eating, on an "industrial scale", salads consisting of vitamin-rich plant-based foods[17] and minimizing fat (e.g., oils), sugar, carbohydrates and even protein. Through this diet, adults consume no more than around 1,600 calories per day (roughly 1/3 of the calories of the average human).

Scientists observing those who have eaten this diet over a long period of time note a more youthful appearance than is seen in those people who consume normal higher-calorie diets. Indeed, individual living on a low-calorie, high-vitamin diet are often described as looking 15 to 20 years younger than they really are.

---

[17] The same vitamins may also be used in facial creams to help provide an appearance of youthfulness in the skin. Minerals may be added to creams to help physically remove dead skin cells when rubbed into the skin, exposing the newly developing skin cells underneath and giving them access to oxygen in the air. Aloe vera and chamomile may be added to repair damaged new skin after the application of minerals, followed by vitamins C and E to protect the cells from free radical damage, all mixed in a plant-based oil such as olive oil or Shea butter.

The only drawback to the low-calorie diet appears to be a slower metabolic rate, resulting in these individuals needing to wear extra clothing to keep warm and sometimes feeling excessively tired if exercise more than usual. The slower metabolic rates of individuals on a low-calorie diet may also affect the rate of cell division (i.e., slow it down) since the amount of protein consumed is probably considered too low to rebuild cells.

Yet, despite these individuals subjecting themselves to a grueling low-calorie diet, their bodies will continue to age. More importantly, their lifespans are not significantly improved through this diet. Now the results of a 10-year experiment from the University of Texas are supporting this view.

Professor Arlan Richardson who headed the experiment got a group of mice with the fastest heart rates (and hence higher metabolisms) together and compared these to other mice with lower heart rates and modifications to a gene to help lower the levels of free radicals. How the mice looked was not of concern to Professor Richardson. One must naturally presume this is because the furry nature of these creatures made it difficult to see how old a mouse was by looking at its facial features. Instead, the focus was on the more quantifiable lifespan of the mice.

In the end, assuming no disease was present, diet remained the same for all the mice, and the mice were sufficiently happy with their environment, Professor Richardson had no choice but to conclude that he saw no obvious difference in the lifespans. In other words, the lifespan was not increasing with a lower metabolic rate and with fewer free radicals.

As Professor Richardson said:

> "When we finally got to the end of the experiment, there was no difference in the lifespan at all, and the genetic manipulation had no affect."[18]

Therefore, the idea that aging is probably caused by excessive free radicals damaging our cells and DNA may only be true in the sense of how we might look at a certain age, but it does not seem to affect our

---

18   Quote from the BBC documentary *Don't Grow Old* (2009).

lifespan. Not even genetic manipulation to reduce the number of free radicals has had an effect.

Clearly something else is controlling the lifespan of living beings—and not even a vitamin-rich and low-calorie diet can completely stop it in its tracks. It must be something deeper and must go at the heart of every living cell. Today, some scientists believe the answer may lie within an important macromolecule known as deoxyribonucleic acid (DNA).

Let's look at this genetic component to the aging process and see whether the answer lies here.

Telomeres

In one important piece of work, scientists noted how special molecules attached to the ends of the DNA called "telomeres" appear to get shorter each time the DNA divides and is rebuilt into two new DNA strands just before cell division takes place[19]. Telomeres are like the ends of a shoelace. They prevent the DNA from fraying and keep the DNA twisted in a tight and rigid structure as needed to resist interference from the external environment. Should the telomeres disappear, the DNA becomes more fragile, and it would appear that the body rejects the DNA, and the cell will not divide. Or if the cell ignores the body's call not to divide, it will turn into a cancerous cell unless the immune system can recognize the cell as needing to be destroyed and removed.

Dr. Bill Andrews of Sierra Sciences is a supporter of the telomere theory for long life. He said:

"Telomere shortening is an absolute problem that we have. Every time our cells divide, our telomeres get shorter. There's nothing we can do about it. No matter how well we eat, no matter how much we exercise, no matter how much we do, or everything our doctors tell us to do, our telomeres still shorten."[20]

---

19   Jack Szostak, Elizabeth H. Blackburn, and Carol W. Greider are credited for their work on understanding telomeres. Their work gained recognition through the 2009 Nobel Prize in Physiology or Medicine. Further details are available in Blackburn and Greider's Scientific American article titled "Telomeres, Telomerase and Cancer".

20   Quote from the BBC documentary Don't Grow Old (2009).

Work is currently taking place to see if it is possible to rebuild the telomeres (i.e., lengthening them) for better restoration of DNA, even though this assumes the genetic code itself lying between the telomeres remains accurate and unchanging over time. As scientists know, the genetic code is the place that stores all the information needed to rebuild any living cell in the body and make them function in the right way.

Of promise in this regard is an enzyme found in reproductive cells that is designed to keep telomeres long. Scientists believe it may even explain why these specific cells appear to defy aging. Whatever the cause, due to the way the enzyme specifically acts on the telomeres and keeps them long, scientists have, naturally enough, called the enzyme a "telomerase".

However, as mentioned earlier, all this assumes the genetic code in DNA is in top condition and nothing interferes with it when the time comes for it to divide and rebuild to form new cells (in order to repair damaged cells from natural wear-and-tear). Should unfavorable mutations take place during this critical dividing and rebuilding stage, or if the DNA is already riddled with some mutations, it is likely the new cells will end an individual's life if enough of them are made, no matter how much the telomeres are lengthened.

Genes

Looking more closely at the genetic code, in an experiment conducted by Seymour Benzer (1921–2007) and his colleagues at the California Institute of Technology it was revealed that a specific gene within the DNA of living cells could extend the lives of fruit flies by as much as 30 percent. The gene, first observed in 1998 and also present in humans, has been dubbed the "Methuselah gene" after Biblical accounts of a man of the same name living to the incredible age of 969 years.[21]

In another landmark study in the 1960s, Dr. Leonard Hayflick, who became director of the Center for Gerontological Studies at the University of Florida, was among the first to suggest that somewhere in

---

21   The results of this study can be found in *New Scientist* published on November 7, 1998 (see https://www.newscientist.com/article/mg16021594-700-methuselah-flies/). Original article where this study first appeared is called "Extended Life-Span and Stress Resistance in the Drosophila Mutant" published in October 1998.

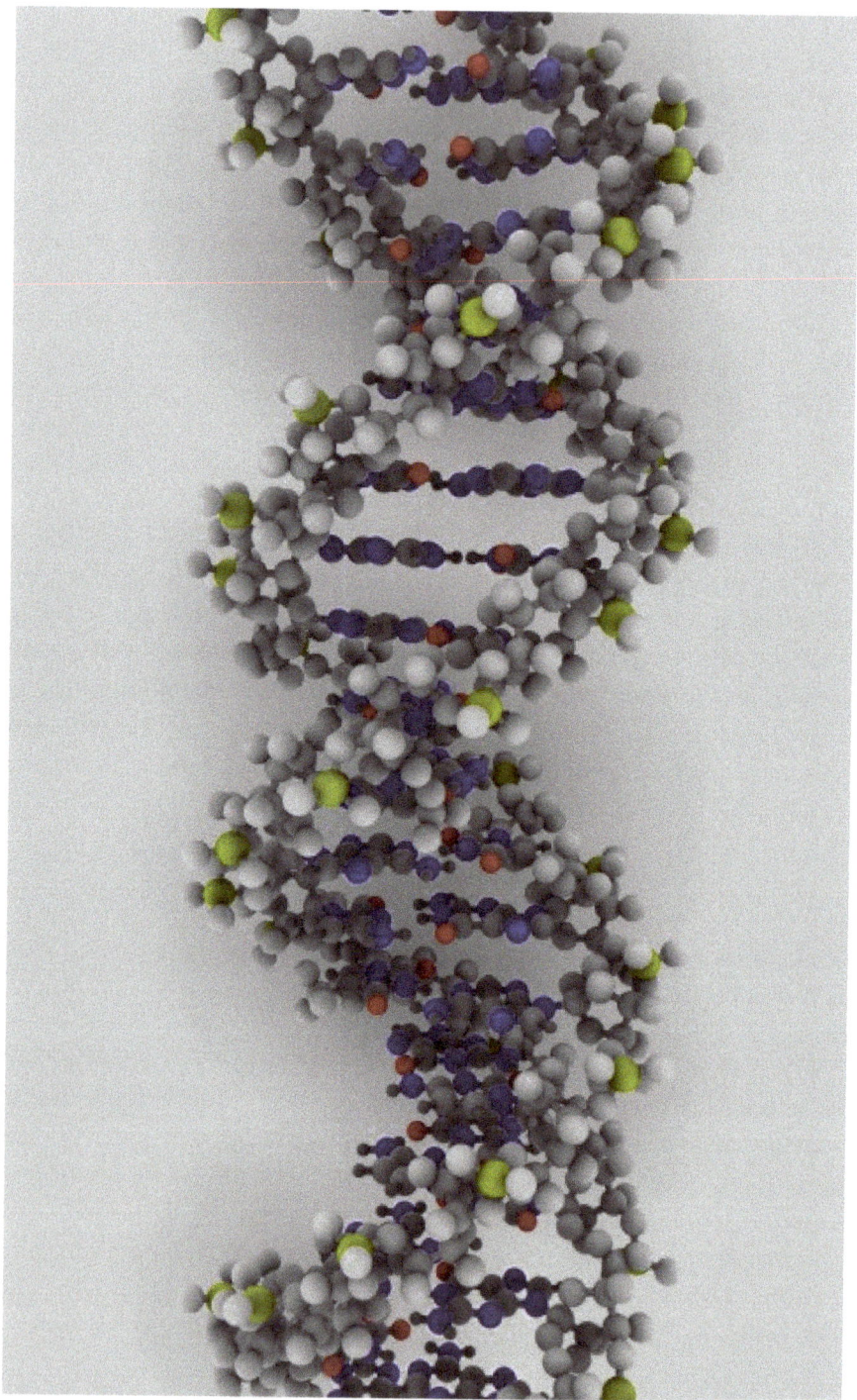

Deoxyribonucleic Acid (DNA), the place for storing the genetic code for all living things.

the genetic code lies the program that controls when our bodies decay. He originally believed that, after the cells in the body reach a certain limit of self-replication known as the "Hayflick limit", a built-in genetic clock would cause the cell to drastically lose its ability to function and eventually cause death to the organism. Hayflick estimated that it would probably take no more than about 50 cellular divisions before a cell collapses.

However, in 1972, Dr. David Harrison, a senior staff scientist at the Jackson Laboratory in Bar Harbor, Maine, performed an experiment to test the Hayflick limit. Harrison used a sample of rapidly reproducing stem cells found in the bone marrow of young experimental mice and transplanted the cells repeatedly within each new generation of mice produced in the laboratory. Each generation of mice had their original bone marrow destroyed by radiation before it was replaced with the transplant material. In conclusion, Harrison found no evidence to suggest that a hidden code in the genetic material is forcing living cells to deteriorate after reaching 50 cell divisions. In other words, there appears to be no known limit to how many times a cell can replicate.

"If Hayflick had been correct," explains Harrison, "when you transplanted stem cells from old mice into young ones you should see a real difference in their functioning after a point. But we didn't see this. Stem cells don't seem to age."[22]

Although the cells showed no signs of deterioration[23], this neither proves nor disproves the possibility that the limit for cellular division may be set at a much higher level. But while the cells remain happily protected inside calcium-rich bones (useful in reducing the amount of radiation from the natural environment that gets inside the bones to damage the cells), it seems they can continuously reproduce themselves indefinitely.

In that case, what happens should the cells be located outside the bones? Do the cells remain unchanged? Or will they die sooner as if they had aged at a quicker rate?

---

22   Batten 1984, p.39.
23   This assumes the genetic code and the telomeres are not changed over time (i.e., there are no mutations in the code, and no shortening of the telomeres). If the cells are grown in culture outside the bone (e.g., in petri dishes), any means to instruct the cells to manufacture telomerase, for instance, to ensure the telomeres remain long after each cell division, as well as any means to prevent mutations in the genetic code, can help these cells to replicate indefinitely with no loss in quality of the cells. Cells, when sufficiently protected (i.e., to prevent changes), do seem to show the signs of immortality.

This is an important point: Because should there be evidence to show that the stem cells do die sooner than expected outside the bones, we have to face the reality that something in the environment has to be interfering with the cells and the DNA itself and preventing them from doing their job properly. The bones—containing the metal calcium—are merely providing better protection from whatever is interfering with the cells.

So, it may not be entirely a genetic cause, but an environmental one as well.

In fact, the idea that cells and DNA do reproduce themselves indefinitely in a protected environment seems to be supported by a particular species of sea creatures known as *scarlet jellyfish* (the scientific name is *Turritopsis dohrnii*). First studied in 1979 by Shin Kubota, a professor at Kyoto University's Seto Marine Biological Laboratory, these 4.5 mm diameter bell-shaped creatures and roughly 5 mm tall (originally from the Caribbean sea and now commonly found in the Mediterranean Sea and in the waters around Japan) have the ability to rejuvenate virtually all of their cells to make the creatures look young again before regrowing back into its sexually mature adult state and can do it as many times as required.

As Kubota said, "They [the jellyfish] don't die, they rejuvenate."

Scarlet jellyfish.

It is strictly not immortality, but it is a way to refreshen the cells of a body through a process known as *transdifferentiation* at the genetic level in order to allow these extraordinary creatures to relive their adult life as often as required so long as diseases, predators and genetic mutations do not impinge on their long-term survival. The rejuvenation usually takes place when the jellyfish is either injured or reaches a certain age. The creature then falls to the ocean floor where it morphs into its smaller infant state, known as a polyp. Over a period of about one month, the cells within the jellyfish are replaced and, after the second month, the jellyfish grows back into a new adult form known as a *medusae*. In one experiment, Kubota has succeeded in rejuvenating one jellyfish 12 times in the laboratory. The question is: How does it achieve this remarkable feat?

After much research, it would appear scientists have managed to isolate specific peptides from the jellyfish called "forkhead box transcription factors" (FOX) in the nucleus of the cells. Closer examination of the function of these peptides have now revealed the biological mechanism for determining whether or not a cell will be renewed. It would appear these peptides only get activated when cell damage is detected. If the damage is relatively minor from factors such as pollution, free radical damage, radiation exposure, and dietary deficiency, the peptides set out to repair the damage by stimulating DNA repair to reverse any evidence of mutations in the genetic code as well as tell the cell when to replace itself with a new rejuvenated copy. However, if the damage is too great and cannot be reversed, the peptides tell the cell to self-destruct. Otherwise, the peptides can act to protect further damage to new cells.

In humans, scientists have found a natural peptide mimicking those found in the jellyfish. It is simply called FOXO3a. Measurements on the amount of this peptide in humans have been remarkably accurate in determining an individual's biological longevity. In other words, those who live to be over a hundred years old have higher peptide levels in their cells.

As of 2014, pharmaceutical companies have mass-produced an artificial version of this human peptide called *Juvefoxo*, and only recently has this substance been added to some skincare creams to help encourage skin cells to regenerate themselves.

Of course, whether these new rejuvenation creams actually work depends on the quality of the DNA in the skins of individuals in the first place. If the DNA has been well-protected from the factors leading to mutations, humans may benefit from the growth of new cells to replace the old. In ocean environments, DNA is certainly better protected than the DNA of animals living on land. Thus, one can see the natural benefit of the peptide to the jellyfish. For humans, however, the verdict is not yet out on its long-term effectiveness as an anti-aging substance.

However, one thing is certain. There is clearly something in the environment to help control the quality of the DNA, which in turn affects how long a living thing can survive. Providing protection from whatever is in the environment seems to make a big difference in the lifespan of living things. The question is, what exactly is in the environment to affect the DNA?

What is in the environment to affect cells?

The term for environmental factors that affect aging is called "gerontogens". Think of these as toxins or foreign bodies in the environment capable of affecting the way living cells and DNA work. This environmental component to the aging process is considered just as important as the genetic component. It is something that must always be looked at as part of a complete study of the aging process.

As Norman E. Sharpless from the University of North Carolina, Chapel Hill, USA, said:

> "The rate of physiological, or molecular, aging differs between individuals in part because of exposure to gerontogens....We believe just as an understanding of carcinogens has informed cancer biology, so will an understanding of gerontogens benefit the study of aging. By identifying and avoiding gerontogens, we will be able to influence aging and life expectancy."[24]

---

24   http://www.sciencedaily.com/releases/2014/05/140528133209.htm.

Let's examine this environmental issue more closely.

We know our environment is littered two main types of dangerous foreign matter:

1. *Large-scale foreign matter*
   Capable of destroying whole organs in the living organism (e.g., a spear or bullet).

2. *Small-scale foreign matter*
   Capable of destroying living cells and DNA (e.g., fungi, viruses, bacteria, chemicals, and radiation).

It is fairly obvious in 1 how the lifespan of an organism can be curtailed by large foreign matter entering the organism's body. So let us focus on 2.

Now, can tiny foreign objects affect the behavior and function of living cells? Yes they can and, in fact, can affect the lifespan of an organism.

Imagine if you threw a foreign object shaped like a tennis ball at someone's face as hard as you could from a distance of, say, 5 meters. You can expect the impact of the ball with the face to create changes. For example, we would observe how the skin and muscles of the face will deform to absorb the energy of the ball. The elastic nature of the skin and muscles may even push the ball away, as the tightly knitted structure of the skin, muscles, and bones is usually enough to absorb the impact.

But throw a cricket ball at someone's face, and the potential for damage is much greater. Bones can break, blood vessels can burst and bleed, and the skin can rupture. All this damage will have to be repaired.

The same is true of living cells.

In other words, the change created after throwing a tiny foreign object at a living cell may be within the acceptable strain limits of the cell. At other times, the interference will be strong enough to cause damage. Should this happen, biological mechanisms within the cells will attempt to repair or rebuild the cells to what they were in order to re-adapt to the environment and continue to protect the living organism once again from whatever is in the environment that interfered with the cells in the first place.

But here lies another important fact we must consider in this aging research: any tiny foreign matter that creates sufficient damage to living cells can affect how well the repair mechanisms can do the job of fixing the cells properly. In other words, the repair mechanisms are not perfect up to a point. Beyond that point, the cells no longer look like the originals that first appeared on the scene through cell division.

It is true. The changes can become too great, and the repairs may not be perfect.

Going back to the cricket ball example, we know that if the cricket ball hits a person's face, there is high probability for structures deeper down to get damaged, such as a fractured cheekbone. True, the bone will do everything it can to repair itself. This is natural. If the damage is too great, however, then not even the natural reparative mechanisms within the bone will do a perfect job. Then, you will notice something different about the face of the person compared with what it was before. Give it more time, and you may also discover how the person behaves and functions differently in the environment as a result (unless you give the person lots of love and care).

Interestingly, the same thing happens to a living cell.

And, as scientists have found, this is exactly what is happening in the aging process. Something deep down within the cells of a living organism has changed sufficiently to make the repair mechanisms imperfect and for the cell to perform its job in a different way than when it first appeared. Should enough of these "different" cells appear to replace the old, it can affect the way the organism looks and performs its activities.

But what has caused this change or damage to the cell?

It may be an invisible foreign object getting in deep enough to cause significant damage to the cells. There can be something in the environment to fool the biological mechanisms within the cells to the point where they cannot produce new cells as exact copies of the original parent cells.

Now, thanks to the Unified Field Theory, we know of one foreign object capable of doing this to a living cell.

If you examine the foreign objects in the Universe, there is only one that affects all living cells at the deepest level and does it progressively over time. Nothing can escape it. Everything else, such as fungi, bacteria and viruses, act too fast and/or tend to be limiting to a certain number of cells—but not to all cells in a living thing or for all living organisms. The foreign matter we are referring to here is relentless and pervasive, and it can penetrate to the deepest levels of every living cell and disrupt certain functions and biochemical reactions. That foreign matter is called "radiation".

Radiation has the power to affect everything. It does not choose; rather, it indiscriminately affects both young and old, different races, and different species. Being female or male makes no difference either. Absolutely nothing can escape the potentially detrimental effects of radiation no matter how hard we try. This is exactly the kind of foreign matter needed to explain the aging process for all living things in the Universe.

Is radiation the answer to the aging process?

Well, it does make sense.

When people talk about ultraviolet light from the sun causing signs of premature wrinkling on the skin, we note it is one of the signs of aging. Similarly, should someone receive unusually high doses of energetic radiation, such as gamma rays, biological structures start to break down, leading to a weakening of the skin and blood vessels, more wrinkles on the skin, sudden bleeding, memory loss, deteriorating eyesight and hearing, and so on. These are, again, classic symptoms of aging.

Even the presence of cancerous cells in the body after exposure to energetic radiation is certainly not unheard of and tends to be of the same type we see in the elderly.

These observations suggest that radiation could be the key to understanding the aging process and hence the lifespan of all living things.

In that case, let us look at the precise mechanism by which radiation intimately affects the living cells, making them age and reducing their lifespans.

The Unified Field Concept to Aging

Understanding how radiation affects living cells

The special macromolecule known as DNA is vital to all living things. If DNA did not exist, the essential biochemical and genetic information needed not only to reproduce your entire body for survival of the human species, but also to run the hardware of the living cells and your entire body in a specific way needed to cope with the environment while you achieve certain goals would be lost. Indeed, you would be no more complex than an atom floating through space. You need DNA to bring together the atoms and form the type of cells and the living body you have today for coping with your constantly changing environment.

The exact task performed by any given cell (and ultimately by the living organism) is determined by various active sections of the genetic code in DNA. These active sections are read by another molecule called ribonucleic acid (RNA), a close chemical relative of DNA, and translated into the appropriate chemical information (i.e., amino acids assembled into proteins) needed to build a new cell as well as direct the new and existing cells to do their job in a specific way. So, if a change is made to an active section of the genetic information of DNA for any reason, it will constitute what scientists call a "mutation" which may alter or completely lose the original function of the cell, and ultimately, of the organism.

Mutations do occur in DNA. Scientists can attest to this fact by examining the differences in the ways all living things look and behave on Earth. All living things on Earth originated from a single living organism that has diversified over billions of years to create the creatures we see today, thanks to the help of mutations. Thus, we owe our very existence on Earth to numerous beneficial mutations carried over through many generations.

However, not all mutations are beneficial.

It is these types of mutation, the ones we call non-beneficial, that scientists believe are one half of the problem of growing old. The other half of the key lies in what is happening in the environment to force DNA to self-replicate in order to rebuild the genetic code and keep it as accurate as possible as well as undergo continuous cell division to help produce a new cell ready to face the constantly changing Universe. According to the Unified Field Theory, the most likely cause for this self-replicating behavior to handle the changes in the environment is that familiar tiny, invisible foreign body called radiation.

Yet, even when the DNA is self-replicating and a new cell is produced, there is a heightened risk of the genetic code experiencing mutations. Let us explain a bit more about what happens during cell division.

First, the DNA undergoes a molecular division, whereby the "double helix" molecular structure of DNA must unwind itself as if "unzipping" into two separate strands. To minimize the risk of damage, however, the entire DNA is never completely unwound. Only a tiny portion of the DNA is unzipped, allowing special protein-like molecules to come in and quickly attach free DNA molecular parts, i.e., nucleotides, which float in the primordial soup of the cell's nucleus to each strand in the correct sequence, resulting in two new DNA molecules. Once the DNA portion is copied, it closes up, and another portion is unwound ready for copying. This process keeps repeating until it reaches the end of the DNA molecule.

After completing the copying process, one of the new DNA molecules is ready to direct other molecules (e.g., amino acids) in the cell to create a new cell, which physically divides from the parent cell in a process known as *mitosis*, while the other DNA molecule stays with the original cell to continue its genetically-assigned job unless the body decides to reject it.

But every now and then, something in the environment will interfere with DNA and its process of self-replication. For the purposes of this research work, this something in the environment can be radiation. Why? Because we know radiation has the particle-like property capable of penetrating the cell to its deepest level, even to the point of interfering with DNA either during, before, or after the copying process. However, the most vulnerable time has to be the moment of copying DNA.

It is at this level of the atoms that a sufficiently energetic ray of light can collide and damage DNA. By damage, we mean breaking apart the

DNA strands or knocking out one or more of the nucleotides composing DNA.

But does this radiation have to come from outside the body? No. It is possible for certain electrically charged sources inside the body to accelerate and decelerate fast enough as they move through the blood stream to create higher-than-expected levels of radiation in the body. These are the aforementioned "free radicals", which can increase radiation inside the body.

Even if these free radicals are completely neutralized, however, we still have the faithful radiation in the environment to contend with. This is a fact of life.

Scientific support for the damaging effect of radiation on DNA is readily available in the literature. For example, the following statement on the damaging effects of radiation on DNA was provided by Christoffer Johansen:

> "...ionizing radiations of much higher frequencies such as x-rays and gamma rays...are genotoxic and known to damage DNA molecules either directly or indirectly through free-radical formation."[25]

Despite the undeniable ability of light (or radiation) to damage DNA, scientists have noticed within the cell certain enzymatic mechanisms capable of regenerating virtually all of what was damaged in the DNA back again. But—and this is the crucial point—as with any repair mechanism, it will depend on the competency of the mechanism in quickly repairing the damage. Now scientists are discovering that no repair mechanism in the cell is perfect.

This is a crucial fact. The repair mechanism we have in our body can be easily fooled if, say, the two strands of DNA are broken simultaneously and one part of the molecule has time to rotate freely by 180° before the repair mechanisms can detect the damage and rejoin the two strands. If this should be the case, we would already have a permanent mutation in the genetic information that may alter the function and purpose of the cell. And, if enough cells are affected in a similar way or the original mutated cell grows through cell division to

---

25   Johansen, Christoffer et al., "Cellular Telephones and Cancer—a Nationwide Cohort Study in Denmark": *Journal of the National Cancer Institute*, Volume 93, Number 3, February 7, 2001, p.203.

replace other cells, the function and purpose of the living organism will change as well.

Of course, natural mutations in DNA created by radiation do not always have to be seen as detrimental to our survival. Occasionally, the change could bring unexpected benefits to the cell and to our way of life by helping us to cope with our environment in a slightly or significantly better way.

However, by the same token, it is true that the changes can also be non-beneficial and can only be weeded out after birth by natural selection (or perhaps by medical science should an effective solution be found). It is here, in the development of non-beneficial mutations in the DNA, where progressive deterioration in the structural quality and function of the cells leads to permanent changes in our appearance and health, such as lowering of the immune system's response to disease, a reduction in flexibility and suppleness of the skin, hardening of the arteries, and a weakening of the muscles, that we see what aging is truly all about.

Should enough of these non-beneficial changes continue to persist with more and more mutations taking place in the DNA over time and with nothing to repair them, they will eventually become more life-threatening to the organism, with the eventual collapse of vital organs or the creation of an incurable cancerous growth that ultimately causes death to the organism.

Support for this "genetic error" accumulation theory as playing an important role in the aging process may be found in the 1992 edition of the Encyclopedia *Britannica*:

> "...[the] genetic theory of aging assumes that cell death is the result of 'errors' introduced in the formation of key proteins, such as enzymes. Slight differences induced in the transmission of information from the deoxyribonucleic acid (DNA) molecules of the chromosomes through ribonucleic (RNA) molecules (the 'messenger' substance) to the proper assembly of the large and complex enzyme molecules could result in a molecule of the enzyme that would not 'work' properly. These so-called error theories have not yet been firmly established, but studies are in progress."

The Secret to Immortality

Assuming radiation plays a significant role in the aging process, is there any way we can prevent this vexatious form of foreign matter from penetrating our body? Or more simply, can we stop the aging process altogether?

According to Einstein's Unified Field Theory, the most important factors in reducing the number of genetic "errors" creeping into DNA caused by radiation are as follows:

1. *Lower the quality (i.e., frequency) of radiation entering your body*
   Low-frequency radiation with its lower energy density will find it harder to knock out atoms from molecules in the body than high-frequency radiation.

2. *Improve the genetic information stored in DNA through beneficial mutations*
   Improvements to your genetic code can produce better proteins and other chemicals to help protect you against the damaging effects of radiation. They may also help to improve the overall efficiency and accuracy of your inherent biological mechanisms for repairing damaged proteins and DNA, not to mention the removal of cellular garbage such as waste products produced from metabolic processes, unrepairable cellular components, and other toxins that are created within, or penetrates into, the body.

3. *Consume and absorb the right types of chemicals into the body through foods and certain types of skin creams*
   Certain foods can provide essential chemicals that help to

neutralize free radicals in the body as well as act as barriers against radiation.

4. *Wear effective clothing to protect against radiation*
Consider wearing skin-tight clothing that contains a fine metal mesh or fiber. This metal will reflect the radiation away from the body. If the radiation still penetrates the body, the symmetrical nature of the body and the skin-tight design of the clothing can make the clothing behave like a Faraday cage by reflecting the radiation at the opposite end back on itself and in an out-of-phase fashion, thereby annihilating the original radiation and dropping the energy to zero within the body.

5. *Stay behind barriers or immerse the body in a solid substance to protect against radiation*
If skin-tight metal clothing is not available, staying indoors behind stone, bricks and/or metal roofs and walls can reduce the amount and quality of the radiation penetrating inside and entering your body.

6. *Lower the mass and density of the body*
A lower mass and density helps the body to absorb less radiation and generates less radiation internally by the body's own accelerating electrons and protons, as well as the quantity of free radicals generated in the body. Thus if you do not have high levels of growth hormones[26] or don't exercise to build muscle that can easily add extra mass to your body and so make you grow bigger, your small size will generally make you live longer.

---

26 The most common natural growth hormones to be found in the body are from a chemical called IGF-1 as well as insulin and glucose. So, if you do things like add protein and/or sugar into your diet, the body responds by thinking it needs to rebuild old cells and create new ones leading to the familiar growth we see in all living things. However, reduce protein and sugar in your foods, and the levels of the natural growth hormones will drop significantly. Then the body goes into what scientists call the "maintenance mode" where efforts to ward off age-related diseases is taken more seriously until there is nothing more to fight off. Then the white blood cells drop in numbers. Furthermore, if the energy levels are low (because you are consuming less sugar), the body also undergoes as part of its maintenance the removal of old cells, or cells that don't help the body in any way, leaving behind the higher-quality cells to do their jobs of keeping the body alive. Should the protein-rich diet be returned, these better-quality cells are what will reproduce to create new cells, ready for the older cells to shed and disappear in the next fasting period of your diet.

The idea of a low-mass body living a longer life is not wholly unheard of in science. In fact, the concept is well-known to astronomers in the study of life cycle of stars. Basically, the larger and more massive is the star at birth, the more energy it must expend, the hotter it gets internally, the more radiation it absorbs and emits, the more fuel it uses up to stay bright and active, and the shorter its lifespan. The same is true in biology.

The lifespan of sea turtles are typically between 30 and 50 years, but some documented cases show they can live as long as 150 years. Why turtles can live this long is not fully understood, but it is likely that prolonged protection from radiation in the ocean environment and relying on the shell on its back to reduce radiation entering the body may play an important role. Other factors may also affect lifespan, including type of diet and living a relatively stress-free life.

As for lowering the temperature of a body (i.e., reducing the frequency of the radiation) to increase its longevity, there is evidence to support this link. According to the 1992 edition of the *Encyclopedia Britannica:*

> "In cold-blooded animals in general, the rate of metabolism that determines the various life processes varies with the temperature to which they are exposed. If aging depends on the expenditure of a fixed amount of vital energy, an idea

232

first proposed in 1908, lifespan will vary tremendously depending on temperature or other external variables that influence lifespan. There is considerable evidence attesting at least to the partial cogency of this argument. So long as a certain range is not exceeded, cold-blooded invertebrates do live longer at low than at high temperatures."

But if you want to slow the aging process dramatically and extend the lifespan to unbelievable timeframes, you have to do some serious work.

Immortality is defined under the Unified Field Theory as the moment when all harmful radiation (the ones that can penetrate living cells and damage DNA) is removed from your body.[27] Barring any accidental mishaps or catastrophes, poor genetic coding in the DNA, a weak immune system, or significant pollution in the environment to add to the foreign matter interfering with your cells, there is nothing in the Universe stopping you from approaching this state. With the right technology, adjustments to the DNA code, and eating the right diet, it is possible to extend your lifespan for literally hundreds, if not thousands of years.

For longer lifespans of hundreds of thousands of years or more, the only way to achieve this is in the field of cryogenics. Here you will find science applying extreme techniques to reduce the amount of radiation penetrating a human body.

Despite the alluring promise of a longer life by reducing your body temperature with the help of science, the inevitable question remains: Can you be revived from the cryogenic state (and will society want to revive you)? And what will the quality of life be like if you achieve extreme levels of longevity? Or will you need to take a more balanced approach and choose a lifespan of perhaps several hundred years instead of thousands or millions of years using alternative low-tech solutions?

You should be aware that, even in cryogenics, immortality is still not yet achievable. The radiation can still penetrate your body, albeit to a lesser degree. Why? Because there are still accelerating electrons and

---

27   Immortality can also occur when the DNA molecular structure itself is the toughest to resisting radiation damage, and the genetic code contains the right information for producing the most robust and perfect repair mechanisms for the DNA and cells as well as protection from much of the radiation in the environment.

protons (i.e., charged particles) in the atoms composing your body capable of generating unwanted radiation. And sometimes, the odd high-frequency radiation in the environment (from stars or the technology employed by people) will penetrate your body. So, you will continue to age very slowly.

For example, if you were to be kept in the deep freeze for nearly 10,000 years, on your return to normal room temperature, you may have aged a year or two. Reduce the temperature further and perhaps the body will only age by a month or two. If you keep reducing the temperature, a point will be reached whereby your body will no longer age at all. We call this the moment where no radiation can be created inside your body or penetrate your body from outside. It is at this precise moment that you will discover the true secret to immortality. True immortality, according to the Unified Field Theory, is the state whereby all electromagnetic radiation capable of being generated and penetrating your body is completely removed, and you are in a state of perfect stability.

This means your body must attain *absolute zero kelvin*.

Yet the question remains: Would you be able to return to room temperature to tell others of your experiences after reaching absolute zero kelvin? Highly unlikely. Why? Because at this temperature, you are effectively creating a perfect vacuum environment of no radiation, and the Universe does not allow for this to happen. It means any solid matter inside this perfect vacuum region will literally explode and pump out all electromagnetic energy composing the electrons and protons in your body (assuming no exotic forces of nature exist to keep matter together at this temperature). Not only will the particles stop accelerating, but the energy making up the particles will probably stop oscillating as well, and then the energy will likely disappear in a great flash of light. In effect, you will turn into the biggest nuclear bomb on Earth.

Look on the bright side: At least a future society might find you useful as an alternative nuclear energy source should your body turn into pure energy.

Until science can say for sure whether or not we can or cannot achieve immortality, we simply do not know precisely what happens to matter when it reaches absolute zero kelvin. No one has seen it occur, and may never be observed. Also, it is not yet entirely clear whether there are exotic forces in subatomic particles to resist this pulling apart of matter in a perfect vacuum. The assumption is that at absolute zero kelvin, all matter will evaporate by turning into pure electromagnetic energy. If the latter is true, you can forget about getting revived. There is no technology in the Universe to bring you back. However, because attaining the coldest temperature known to science is an impossibility, your body put in the deep freeze will always age over time. Nothing can stop the aging process.

Until scientists find the answers, it might be better for everyone to stop dreaming about immortality and just learn to live and experience life as we have been given in this moment in time. Sure, there is nothing stopping you from doing things to extend your lifespan. Just looking around us, we see evidence of people living longer and longer because of better hygiene and cleaner and higher-quality foods. Even the type of clothes we wear can make a big difference. And if your stay indoors a lot and/or use metal clothing to cover your entire body, you will reduce a significant amount of the harmful radiation, including ultraviolet light, x-rays and a certain amount of gamma rays. But again, such protection is not perfect. Radiation will still penetrate the clothing and buildings. All it takes is one exceptionally energetic gamma ray to pass right through brick walls and metal clothing and cause irreparable damage to DNA in your living cells. Still, if you can handle this and keep a smile on your face, then the potential is there to live hundreds of years. Or perhaps you are happy to live to 130 years by eating the right diet and being happy and relaxed? We see this is common in parts of the Japanese population and people living in the Mediterranean coasts with their consumption levels of vitamins and minerals, natural vegetables, a glass of quality wine, virgin olive oil, and a bit of fresh protein. Remove the stress of life and people have been known to consistently live to well over 100 years.

So, imagine if people start focusing more on radiation and finding means of protecting themselves from it, as well as eating the right foods and being happy. Perhaps metallic clothing, to reflect radiation, will become a new fashion accessory in the future; wearing it may extend our lifespans by decades or centuries. Whatever we do, however, we all have to strike a balance.

In the meantime, scientists will undoubtedly continue to learn more about the aging process. As our scientific knowledge expands, so will our lifespans, and probably, so will the length of time we are able to maintain our youthful appearances. Don't be surprised if, say, a hundred years from now, people live quite happily for two or three hundred years or even longer (and look young well into their old age), just by reducing the amount of high-frequency radiation they receive every day from the environment, such as wearing metallic clothing on a regular basis. This is certainly not beyond the scope of science or anyone living in the real world to achieve.

As Benjamin S. Frank, M.D. said:

> "It is apparent that aging itself is now a treatable [not curable] disease. A lifespan of more than a few centuries will, before long, become a potential reality."

Until then, the knowledge of how we can live a very long life and look young may now be within our reach, thanks to the Unified Field Theory.

CHAPTER 8

The Link Between God and Light

Science without religion is lame. Religion without science is blind.

—Albert Einstein[1]

WE HAVE reached the final immutable aim of Einstein—to understand God (or at least the "mind of God"). His interest in a thing called God is evident from various statements he made publicly, such as,

I am convinced that He [God] does not play dice.

God is subtle but he is not malicious.

---

1   Originally appeared in Einstein's article titled "Religion and Science" published in *The New York Times Magazine*, November 9, 1930, pp.1-4. It has been reprinted in *Ideas and Opinions*, Crown Publishers, Inc. 1954, pp.36–40, and in Einstein's book *The World as I See It*, Philosophical Library, New York, 1949, pp. 24-28.

God does not care about our mathematical difficulties. He integrates empirically.

I cannot imagine a God who rewards and punishes the objects of his creation and is but a reflection of human frailty.[2]

I want to know God's thoughts; the rest are details.

These examples offer a rather remarkable display of candor on a controversial topic considered more attuned to religion than science. Even recently, a poll conducted in 1998 revealed that as many as 93 percent of National Academy of Sciences (NAS) members, one of the most elite scientific organizations in the United States, did not believe in God[3]. Of course, the NAS represents only a smidgen of the entire scientific community with just 2,000 scientists. Other scientists are more likely to be a little more open-minded. In history, we know that Max Planck (1858–1947), Charles Darwin (1809–1882) and Michael Faraday (1791–1867) were among those scientific elites to believe in God. For example, Faraday gave hints of his belief in God when he said:

> "The book of nature which we have to read is written by the finger of God."

Whether this is a justified position to take as a scientist is unclear, but it has certainly not hampered the quality of Faraday's scientific work and those of his contemporaries. And, at any rate, this chapter will attempt to find out if these scientists are right in their interesting and personal convictions.

Luckily for this chapter, Einstein himself was also open-minded on the God issue. We learn that his understanding of God started at a very young age thanks to his religious upbringing on his mother's side. A very well-educated, quiet, and precocious woman, Pauline Einstein not only made her son Albert learn to play the violin due to her inclination toward the arts, but also instilled into him a number of ideas from her own Jewish upbringing, including a firm belief in the existence of God.

2   http://www.articlesbase.com/meditation-articles/albert-einstein-said-i-cannot-imagine-a-god-2482749.html.
3   http://www.stephenjaygould.org/ctrl/news/file002.html

That upbringing helps explain why Einstein felt comfortable with the issue to the point where he probably wanted to know who or what God is.

The other likely reason for Einstein's comfort is that he may have personally acknowledged a oneness of the Universe after completing his unifying attempts to bring together a number of scientific concepts. This work could have made it easier for Einstein to see and understand this mysterious oneness of the Universe, and so better see "the mind of God".

Whatever the case, Einstein seemed determined to find out the true nature of God through his own scientific work.

But does God exist? What is God precisely? And did Einstein believe in God? And if God does exist, did Einstein see "the mind of God" to help him fully understand the Universe and what it is doing, and why we are here?

To find possible answers to these questions, we will begin this quest by gathering some of the descriptions of God and words used by the religious leaders of the world. Since they have spent a great deal of time contemplating and discussing the existence of God, it would seem reasonable to ask such leaders, "How do you explain God in ordinary language?"

The Names Given to God

The first thing we learn from the religious leaders is that God has many different names. The tendency to assign a name or names to this mysterious entity both verbally and in written form is particularly common among those living in highly communicative civilizations comprising various languages and cultures.

For example, in German, we find God can be called *Gott*; in Scandinavian, *gud*; in Tamil, *Kadavul*; in Persian, *khoda* or *khuda*; and in Vedic Sanskrit, *hu-* (meaning "to invoke the gods"), to mention just a few.

In Judaism and Christianity, the terms "our Father", "Him", "He", "God" and "Lord" are used interchangeably. Furthermore, the English translation of the Holy Bible also makes a distinction between the capitalized version of the word "God" and "god"[4] in an attempt to show that all living things (i.e., gods) capable of influencing all other living things must strive to improve (i.e., change) if they wish to properly understand the principle of love based on the feedback of those being influenced. It is through these changes that people get closer to seeing the ultimate Truth or knowing what it is like to be "God".

For example, in John 10:34 of the Bible, Jesus of Nazareth (ca. 6 BCE–CE 33), also known as Jesus Christ, asked the following to a group of skeptical Jewish men in Solomon's temple, "Is it not written in the scriptures that you are gods?"

In asking this, Jesus believed that we are all "gods", including himself[5], in the sense that we have the power to influence and change things. Whether this is done in a positive (or God's) way, especially when likely to affect living things (i.e., those with emotions), depends on our understanding of the principle of love[6]. Because we are not perfect creatures (i.e., not God), the principle of love is something we must strive to understand and will continue learning throughout life. Learning alone will not make us immediately perfect, but it will help us to improve our understanding of love to help minimize any harm (or

---

4   A god is also called a "deity". For God, the first letter only needs to be capitalized to have the same meaning i.e., "the Deity".

5   The assumption here is that Jesus is not God. Christians religious leaders, on the other hand, are more inclined to believe Jesus was God. Whether that assumption is true or not may depend on the manner in which Jesus came into the world through a human mother named Mary. Among many Christians, there have been talk of Mary being a virgin all of her life yet the Bible claims she could not understand how she could get pregnant when a mysterious angel approached her and mentioned what was about to happen. Because of the mysterious nature of her pregnancy, it is thought by Christian leaders that this is evidence of God as he allegedly entered the world through natural child birth. Then again, it could also be true that a sophisticated god that arrived to Earth in the past and capable of influencing human affairs through its actions had the ability to artificially impregnate an unconscious Mary. In this latter scenario, it is not necessary to invoke the idea of God or to believe that God appeared in human form. Just to add to the difficulty in accepting perfection in one man, we learn from the Bible that this apparently perfect man expects others to do greater things than he did. As Jesus allegedly said from John 14:12 (NKJV):

"Most assuredly, I say to you, he who believes in Me, the works that I do he will do also; and greater works than these he will do, because I go to My Father."

Surely, it is impossible to imagine another person somehow doing more perfect things than Jesus did if one were to accept the Christian leaders view on the perfection of this one man in history. In that case, who is the real biological father to Jesus? We can only speculate on the numerous, potentially amusing, and yet interesting, beliefs Christians have created over the ages.

6   The principle of love involves finding ways to achieve something of benefit to all living things without causing harm. The true principle of love requires the solutions to be unconditional. Of course, as imperfect creatures, we will make mistakes. But if we are prepared to learn and improve ourselves, then we are following the path "with a heart", or the path of love that the true God expects all living things to attain.

potential harm) to all those who will be influenced by our actions. Ideally, the ultimate aim is complete elimination of harm. But before we get to this position of being like God in a perfect loving way, we should naturally solicit feedback from those who are, or likely to be, influenced or affected by our work. Such feedback would tell us whether our actions are positive and helpful; if they are, that would show we are on the right track. We would have learned to become much better and more loving, getting us closer to an understanding of God through our work.

However, a number of Jewish leaders and their followers in the time Jesus was alive did not grasp the fundamental difference between the concepts of "god" and "God". The problem was further exacerbated by the fact that the preferred Hebrew word for "God" or "god" lacked any distinction. To elucidate this point, let us look at the original ancient Aramaic word for "God" prior to 700 BCE.

According to biblical scholars who have studied a number of ancient texts written in the Middle East, the earliest Aramaic word for "God" was[7],

To the linguistic expert, this "word" comprises two Aramaic "letters" combined with several dots positioned in such a way so as to have meaning and to help those who spoke the language in visualizing the picture behind the "word". Essentially, the three dots at the top of the symbol represent "the Father", "the Son", and "the Holy Spirit". Another way of visualizing this term is to imagine the dot at the top as representing God and the two dots directly below can be thought of as representing the opposites of the Universe. A classic example of opposites include the body (the means by which all living things can affect the physical nature of the Universe and so achieve certain worthwhile goals) and the spirit (this is the everlasting "thing"

---

7   Early Aramaic word for God obtained from https://origin-api.ning.com/files/QFSjLPszBbkUaVk3fMlLp71GccN8rrmSN4bDA0wZ4nWQakJutQ-MZEYTpY*G9XWQCHxIl6uhL4jFKDcRyzpbop7tlmlHlQu9/pendentcardflat.jpg.

representing our true self that is unaffected by whatever happens to the body).

Below the three dots are two important Aramaic letters telling the Aramaic-speaking people how to pronounce the word for God. The only slight problem we have is that no one knows precisely how the combination of these two letters would have sounded more than 2,800 years ago. The best linguistic scholars can do today is compare this Aramaic word to similar words in other languages (especially the Hebrew language, which is known to have borrowed a number of the pronunciations from the Aramaic-speaking people, and the modern Aramaic spoken by a few communities in the Middle East). Today, these experts have suggested that the Aramaic letters may have been pronounced independently from left-to-right as either "yodh-heh" or "yuth-hay". However, in the Aramaic language, the sounds of each letter must be combined appropriately. Thus, it is believed by linguistic experts that the original word for God in Aramaic was likely pronounced as "Yah".

As for the solitary dot sitting below the two Aramaic letters, it represents the "one universal God".

This Aramaic symbol for God was often employed by ancient Middle Eastern scribes to indicate its highly sacred nature, usually because it contained some kind of knowledge pertaining to God. That knowledge came either through stories handed down through the generations or through the writer's real-life experiences (perhaps after an encounter with God).

Then Jesus learned the Hebrew language and listened to some of the teachings from Jewish priests in the stone temples. Given his own perceptive look at the social issues of the day, Jesus noticed a general reluctance among these Jewish religious leaders[8] to get involved in certain social issues or to fully understand the primary causes behind those social issues. Not only that, but the leaders felt hesitant to find appropriate solutions and influence the Romans and others to create a

---

8   In the town of Capernaum, where Jesus lived and began his preaching, a number of Jewish priests held a large and prestigious-looking stone temple built at the request of a Roman Centurion stationed in the town. The temple had a plentiful supply of was fresh running water controlled by the priests (and provided to the Romans for their bathhouses). Jesus saw this situation in the face of certain health problems in people within the community. As a religious leader himself, he engaged in an act of defiance against the religious authority by using the water during the night (including the Sabbath when the priests were not around to see) to help clean people who were being shunned for their sickness as punishment from God. The cleansing was a means of healing various common human diseases prevalent at the time.

better way for people to live harmoniously. In a sense, the Jewish leaders didn't want to rock the boat and chose not to be a "god" in changing the situation for the better.

The reason for this inaction can be understood when we look at how the Jewish leaders were understanding the concepts of God and god through language. Here we see the Jewish men were complicating the issue of God far more than it was necessary by creating and supporting various Hebrew words representing "God" (or "god") and using them interchangeably with no obvious way of distinguishing the terms. By doing so, the Jewish leaders thought they were getting closer to God through their discussions and choice of words. But where they gained certain knowledge in one area, they started to lose knowledge in another. The difference between "God" and "god" was one piece of knowledge they had forgotten. The leaders, therefore, believed they did not need to do anything to be "gods" to help other people, especially the non-Jewish ones, claiming God will one day solve all these social problems for the people, and if not, it could be assumed that God was punishing them.

To help see the complications taking place in the Hebrew language, let us compare the words used to describe "God" and "god" in Aramaic with the Hebrew equivalent.

Before Jesus noticed various names and titles given to God (or god) in the Hebrew language, his primary language as a young boy was Aramaic. Not quite the original early form of the language prior to 700 BCE when the original ancient symbol for God was created. As Biblical scholars can surmise from a study of various surviving fragments of Aramaic texts written at different times, some important changes were made to the language after 200 BCE, partly because Hebrew—another language spoken in those times—had influenced the more popular Aramaic language. Despite the changes, the Aramaic-speaking population, including Jesus, appeared to have spoken of God using one of the two following words:

Elah (perhaps pronounced as *Eloh* or *Elaha*)
> A singular noun representing "God". The word had likely originated from the root words "fear" or "reverence", and thus may be translated into English as "the awesome One" or "the

fearful One". The word is used in the Aramaic portions of the book of Ezra and Daniel, and in one verse in Jeremiah 10:11.

Elahin

A plural noun of *Elah*, denoting strictly "gods". Never is the word used to denote "God" in any of the Aramaic portions of the Old Testament. The word is used at least once in Daniel 5:23.

In Hebrew, the Aramaic word *Elah* can be written as אלה. Alternatively, it may be written as האל according to Google Translate. In other online sources, the simplest Hebrew word for "God" is אל (simply translated as *El*).

The Hebrew word *El* is actually one of the oldest Semitic words for God used by the Jews. The term is mentioned in the Hebrew Bible only occasionally (e.g., Genesis 33:20 and 46:3), but usually in combination with other words to help emphasize some specific attribute of God, such as *El Elyon* ("Most High God") and *El Olam* ("Everlasting God"). If not combined with other words, then on its own, *El* simply means "might, strength, or power", so it could be translated into English as "the mighty One", "the strong One", or "the powerful One". Whatever the translation, this word is considered the closest in meaning to the Aramaic word *Elah* (or even the original *Yah*). In fact, the Aramaic people may have created a new hybrid word from *El* and *Yah* to create *Elah*. The Jewish community perhaps looked at it and thought it was such a great idea that they initially accepted this modernized Aramaic word for "God".

But as time passed, the Hebrew language became more sophisticated with word choices. Soon, other words were introduced to describe "God" or "god". One such word is *Baali*, meaning "my lord" or "my master", except some people disagreed on this word choice. In the Book of Hosea, the prophet Hosea criticized Jews for using this word to represent God, mainly because it sounded too similar to the word *Baal*, which was being applied as the title for virtually all gods by way of statues or beasts in various cities. That similarity, Hosea argued, made it difficult for simple-minded folks to see there is only one true God. Instead, Hosea recommended the title *Ishi*, which means "my husband" as a better term for "God". Here is the verse supporting this word choice in Hosea 2:16:

> " 'It will come about in that day,' declares the LORD, 'That you will call me *Ishi* and will no longer call Me *Baali*.' "

Then, the original word *El* for "God" in Hebrew language started to have a variety of different words added to it to give different meanings and descriptions for "God". For example, here are the modernized Hebrew words used for either "God" or "god":

Eloah (possibly pronounced as *Elowah*)
> The singular noun of "God" or "god". The word is used in the Bible approximately 57 times, most heavily in the Book of Job (about 15 times).

Elohim (or *Elohay*)
> The plural noun to represent both "God" (e.g., Exodus 32:1; Genesis 31:30, 32; etc) and "gods" (e.g., Exodus 9:1, 12:12, 20:3 etc). The word is used 2,600 times in the Hebrew Bible (Tanakh, Old Testament).[9]

Thus, we know the singular Hebrew word for either God or god is אלוה (pronounced as *Eloh*, *Eloah*, or *Elah*). Then we have the plural form of *Elah* (Aramaic) or *Eloah* (Hebrew), which is the widely used Hebrew word אלוהים or sometimes spelled אלהים (pronounced *Elohim*). All of these words have a common link to the original word אל (i.e., *El*) as if God exists in both plural and singular words. For example, אל (or *El*) is subsumed into אלאהא (pronounced *Elaha*, or *Alaha* in Aramaic Syriac as used by the Assyrian Church). Today, many religious scholars believe Jesus had used *Elaha* to mention the one true "God" in his own teachings in both Aramaic and when he spoke in Hebrew to the Jewish scholars.

However, Jesus noticed something else. He could see how the Jewish priests (who in themselves were also religious scholars) were quite

---

9 The only surviving New Testament documents are written in Greek. Here, God is always written in the masculine and singular form known as "theos". Likewise, the word "Lord" to represent "God" is always singular and appears as "kurios". There is no way to distinguish plural and singular forms of the word "god" in Greek.

happy to create their own knowledge of God by uncovering a plethora of other words to represent the same entity, such as *El Shaddai* ("God almighty", "mighty God" or "God the all-sufficient One") and *Elyon* ("highest").

Not to mention, the Jewish men continued merrily adding more words to represent and describe God. For example, the Aramaic word *Elah* (or *Eloh*) can be used as a common word to describe God in Hebrew as follows:

Elah (meaning "the Awesome One", or simply "God")
Elah Avahati (meaning "God of my fathers" e.g., Daniel 2:23)
Elah Elahin (meaning "God of gods" e.g., Daniel 2:47)
Elah Yerushelem (meaning "God of Jerusalem" e.g., Ezra 7:19)
Elah Yisrael (meaning "God of Israel" e.g., Ezra 5:1)
Elah Shemaiya (meaning "God of Heaven" e.g., Ezra 7:23)
Elahi, Elohi, Eloi (meaning "My God")

Alternatively, the Aramaic word *Elah* can be replaced with the Hebrew words *El, Elohim,* or *Elohay* and the meaning remains the same. Thus, all of the following become possible,

El Elyon (meaning "The most high God")
El Olam (meaning "Everlasting God")
El Roi (meaning "God who sees me")
El berith (meaning "Covenantal God")
Eli (meaning "My God")

Or,

Elohim Kedoshim (meaning "Holy God")
Elohim Chaiyim (meaning "Living God")
Elohay Selichot (meaning "God of Forgiveness")
Elohay Elohim (meaning "God of Gods")

And the above list does not even include the original Hebrew word that translates as "Lord" known as "YHVH", which is mentioned approximately 7,000 times in the Old Testament. For example, the following terms appear:

YHVH Elohim (meaning "LORD God")
YHVH O'saynu (meaning "The LORD our Maker")
YHVH Shalom (meaning "The LORD of Peace")
YHVH M'kadesh (meaning "The LORD who makes Holy")

Yet the Jewish men did not stop there. These men quickly moved away from the *El, Elah, Elohim, Elohay,* or *YHVH* patterns, finding a range of new and different Hebrew words for this mysterious entity without regard to any link to the original and simplest Hebrew word for God. Thus, it is not uncommon to link, for instance, for "God" or "god" to be linked to the Hebrew words for "master" (pronounced as *Adonai* or *Adonay*), "lord" (pronounced *adon*), and "our father" (pronounced *Avinu*), to name just a few examples.

Why does the Hebrew language use various terms for what should be the same mysterious entity lying at the heart of all religions? It appears to be because religious Jews wanted to create many of the words as an epithet highlighting the different roles or characteristics allegedly seen in God. Well, if that does not explain it, then the Jews were at serious risk of returning to the dangerous old pagan game of introducing multiple deities into society, initially through the use of plural words suggesting there are multiple "Gods". And then later, more simple-minded folks might fall into the trap of creating statues to represent their own understandings of God from the different words created and start idolizing multiple gods without realizing there is only one God.

This problem is exactly what Jesus saw. Apparently, the Jewish priests were using the Hebrew word *Elohim* in both the singular and plural form to represent God to the point where they could not see a distinction between "God" and "god" (a bit like the earlier situation with *Baali* and *Baal*). Furthermore, the word *Elohim* was being used to replace the older and more appropriate word *El*, which represents a singular God. Not even *Eloah*, with its emphasis on the singular form would be used enough when representing the one and only God in the Hebrew language. It is here where Jesus saw the biggest problem. How could anyone see an obvious distinction between "God" and "god" (or "gods") when using the word *Elohim*?

It soon became clear to Jesus that none of the words helped him or anyone else to relate to this singular God and the purpose of "gods" and who they represent in solving world problems. This realization helps explain what happened next when Jesus decided to challenge the religious knowledge held by the Jewish religious elites. As the story goes (John 10:22–39), the Jewish men thought Jesus was being blasphemous by using the word "God" (when speaking in Hebrew to the Jewish scholars and priests) when he explained to the men that, "I [Jesus] and [my] Father are one". Such a concept was clearly a controversial approach, one sure to shake up the religious authorities of the day, but Jesus simply wanted to do was to test the Jewish leaders' understanding of the concept of God through his carefully chosen words. He was ultimately trying to see whether the men would take his words literally or look deeper into the meaning. But all Jesus was trying to say was that his life's work is what God would have expected him to do. Should his work follow the principle of love as God commanded, Jesus would then know he was aligned with God in this way. So, in a sense, he was one with God.

Jesus wasn't saying he himself was God. Far from it. Rather, it was only what he did through his life's work that made him as "one" with God. The Jewish religious leaders misunderstood this dynamic, so Jesus had to clarify to the Jewish men by stating, "you are gods" in the statement:

> "Is it not written in the Scriptures that you are gods?" (John 10:34)

while being clear in his own mind of the distinction between "god" and "God" (or "our Father"). Indeed, one can conclude that the writer of this gospel, named John, understood that distinction as well (or the English translator of John's gospel made the correct distinction independently).

To put it simply, there is only one God, but there can be many gods. We are ourselves among those gods because of our ability to influence and change things. That ability comes from each of us having the power to act like God when influencing the things around us. More specifically, we all have our own different views, perspectives, and belief systems of how the world should be, and each of us has unique tools to shape our own environments and the thoughts of others. Yet, no matter what we do to change the world (hopefully for the better as a way of getting closer to God), we are effectively aligning with God through our work. If, for any reason, it isn't (through the feedback of others), we must be prepared to learn how to be a better person before we can have an inkling of the one true God and, therefore, know whether the work we are doing in the world is aligned with God. Or as Jesus put it:

> "...believe the works, so that you may know and understand that the Father is in Me, and I in the Father." (John 10:38)

Unfortunately for Jesus, the Jews simply did not understand. And as a result, Jesus ended up being persecuted and eventually crucified on the cross by the Romans at the request of the Jewish priests. John 5:16–18 explains why:

> "And this was why the Jews persecuted Jesus, because he did this on the sabbath. But Jesus answered them, 'My Father is working still, and I am working.' This was why the Jews sought all the more to kill him, because he not only broke the sabbath but also called God his Father, making himself equal with God."

Alright. So now having finally mastered the distinction between "god" and "God", we quickly see a menagerie of capitalized and alternative names for God from various offshoots of Christianity and other world religions. Here is a sampling of the long list of names: *the Grand Creator* (Christian), *the Lord* (Christian), *Jehovah* (Mormons), *Yahweh* (original Hebrew), *Allah* (Islam[10]), *the Truth* (Christian), *our Father* (Mormons and Christians), *the Ultimate* (Christian), *Jah* (found in the

---

10   The Arabic word "Islam" literally means "submission". In other words, do you submit to God by opening your mind, searching for the unity of all things, and acknowledging the oneness of the Universe? If so, you are a Muslim. In fact, anyone with an interest in God and who sees God in all things can be called a Muslim. Christians and Jews. All religious people are there to search for a better understanding of God and believing in his existence are effectively Muslims. Even Adam and Eve in the Garden of Eden, in accepting God, are very much Muslims. It does not matter what term is used to describe a religious person, they all should be part of the one religion of God. All world religions are merely making a representation of this one religion. But in order to see the ultimate religion with greater clarity, it is necessary for people to unify all world religions, including science (if it has anything to say about the nature of God).

Rastafari movement), *the One* (in the United Church of Canada), *Ishvara* (Hindu), *Anami Purush* (meaning "nameless power" in Surat Shabda Yoga), *Akal* (meaning "The Eternal" among the Sikhs), *Ahura Mazda* (Zoroastrians), and *Igzľabihier* (meaning "Lord of the Universe" in the Ethiopian Orthodox Church). As Hans Küng noted in his book *Does God Exist?:*

"The God of all religions has many names."[11]

It is the capitalization of the first letter in the word for God that is understood today as the crucial step in making the distinction by people between God and ourselves (or anything that influences others). This simple modification helps us see all these words as the one and only God. Any other term used to describe a god must be kept in the lower-case form.

However, there is a problem with creating multiple capitalized words for God and lower case words for "gods". Speaking the words can be confusing. If they sound the same, will people know the difference? Even if we choose different words, we now have an almost unlimited number of sounds people can use to represent God. If we are not careful, it is possible for people to inadvertently choose a certain sound to represent their understanding of God, only to be misinterpreted in the wrong way by other people with a different language and choice of sound for representing God. Who knows? Perhaps the probability of wars starting could be very high because one person spoke of God using a particular sound pattern that might mean the worse four letter word in the English language. How do you know the word and sound you have chosen is ultimately the correct one for God? Can anyone know for sure?

This lack of certainty and agreement may explain why no names or sounds are used for God in some religions

For example, in Eastern mysticism, religious leaders (or "mystics") fully appreciate the difficulties people have in describing God precisely through words. Such leaders say that words (or even pictures and statues created by our hands) and the sounds we produce to represent this entity are simply not adequate to describe God with any precision. In fact, the words and sounds used in human language are so imprecise

---

11   Vardey 1995, p.712.

that many mystics have chosen not give a word or sound to describe or name this mysterious entity. For mystics, God must be experienced, not described.

Here, mystics would prefer to experience the closest possible thing to God by practicing quiet meditation for extended periods of time or sometimes taking long walks through the mountains. Such practices allow mystics to question the very essence of God based on their more Eastern religious perspective.

This is an interesting point. If we go back to the original story of Moses handed down through the generations and recorded in the Hebrew Bible (or the Old Testament in the Holy Bible of Christianity), God has no name. In Exodus 3:14, Moses asked God for a name, and he allegedly received the following reply in Hebrew: "ehyeh esher ehyeh". When translated to English, the phrase literally means, "I am what I am", indicating that God did not wish to reveal his name. Of course, this statement does not mean that God was incapable of suggesting a name to Moses by giving some specific sound pattern for humans to pronounce and a symbol to write down. In fact, after Noah's encounter with God, humans reportedly did exactly that, by creating the sound "Yah" and its corresponding ancient Aramaic symbol to represent God. Later, the Jewish scholars made their own contribution by extracting "yeh" from God's response to Moses and forming the word "Yahweh" (or YHWH). However, God deliberately chose not to name himself—not because he could not speak the language spoken by Moses and create a suggested sound pattern and symbol, but because God, quite rightfully, wished to remain nameless. There really should not be a name. This seems to be reinforced elsewhere in the Bible in another statement allegedly made by God in Exodus 23:13 (King James Version):

> "And in all things that I have said unto you be circumspect: and make no mention of the name of other gods, neither let it be heard out of thy mouth."

Apart from preventing people from worshipping other gods once they are categorized with a name (and sound), this situation suggests that the one true God actually has no name (or it is unmentionable, or should not be spoken, but probably experienced and learned from gathering knowledge). Such namelessness ensures all gods spoken and/or created by the hands of humans are nothing more than a representation of God, but not the true God. In order to see the true God (and avoid religious conflict with different groups of people), it is necessary to unify everything we see and hear. Not using words and sounds is a first step. The second step in seeing the oneness of God is felt through experience rather than through knowledge. Does this mean the Eastern mystics are closer to the truth in their understanding of this mysterious entity?

Paul Collins, the author of *Judgment Day – The Struggle for Life on Earth*, supported this view when he said the following on ABC Radio National's *Encounter* program:

> "The most extraordinary thing that I find is that talking to many people, especially to bushwalkers, they talk about experiencing a kind of presence out in the wilderness, a kind of a haunting sense of something that's there, but is intangible, can't be pinned down, you can't put a name on it, but there is this sense that there is something personal there. What I always find interesting is that is exactly what all of the great mystics, Christian mystics, Islamic mystics, Jewish mystics, Hindu mystics, all talk about the divine and the transcendent, in terms of presence, and they are all hesitant to put a name on it."[12]

After analyzing the names used by people to describe this one and only God, one thing seems certain: religious leaders are suggesting God can be nameable or not nameable.

Is God Feminine or Masculine?

More information is clearly needed to help us identify what this entity is in reality. As we have not yet identified who or what God is, let us approach this problem differently by considering whether God has a specific gender. In other words, is God male or female? Or does God show characteristics that we could acknowledge as masculine or feminine?

---

12   ABC Radio National January 30, 2011. http://www.abc.net.au/m/enounter/stories/2011/3120525.htm.

Interestingly, religious leaders seem to have opposite views on the answer to this question. For example, Western religious leaders (i.e., Christians and Jews) see God as a being of absolute unity whose gender is described as a man (to support the terms "Him" or "our Father" used on a regular basis in the Bible), who is old (to support the view that God has wisdom), and who sits on a great throne (to support the idea that God is an absolute authority).

Even in Western art, there is considerable effort among artists to paint God in this

Zeus, the Greek God.

distinctly masculine manner. It is as if the artists rely entirely on the seemingly accurate descriptions of God by many Western religious leaders for inspiration.

For example, a masculine God is clearly captured by British artist William Blake (1757–1827) in his painting *The Ancient of Days.*

Italian painter and architect Raffaello Sanzio (1483–1520) supported this masculine view in his painting known as *Ezekiel's Vision of God.*

Christian artist Luca Giordano (1634–1705) also rendered his own vision of a vividly masculine God in his painting *The Dream of Solomon.* In this story, God provided wisdom to Solomon as he fell into a deep, spiritual sleep.

Of course, the most famous image that most likely comes to mind is Michelangelo's graphic depiction of God in the mural on the ceiling of the Sistine Chapel.

These unmistakably masculine images of God are not the exclusive domain of Christianity. In both Greek and Roman mythology, there are numerous exquisitely made sculptures dating back to the 5th century

The Dream of Solomon, c.1693 (oil on canvas) by Luca Giordano (1634–1705). Prado, Madrid/The Bridgeman Art Library.

BCE showing Zeus, the king of the gods (and all goddesses), looking rather masculine in anyone's language.[13]

Looking at these examples, they understandably lead us to suggest that either God is genuinely masculine, or these paintings and sculptures of God were heavily influenced by the artists themselves, who happened to be male. Or were these great artists of their day influenced by the Western religious leaders who believed God had to be a male?

Perhaps we can't blame the religious leaders. In the Bible, we find numerous stories of encounters between humans and God. Whether or not such encounters actually transpired, something in the way God spoke to humans suggests that God was masculine. For example, Moses claimed to have heard God speak to him. Acts 7:30–32 stated the following:

> "After forty years had passed, an angel appeared to him in the wilderness of Mount Sinai, in the flame of a burning thorn bush. When Moses saw it, he marveled at the sight; and as he approached to look more closely, there came the voice of the LORD."

Moses made no mention of the voice being feminine when he spoke to his followers about his encounter. Even when God decided to show itself to Moses, God was still seen as masculine. As Numbers 12:4–15 claims:

> "Suddenly the LORD said to Moses and Aaron and to Miriam, 'You three come out to the tent of meeting'. So the three of them came out.
>
> Then the LORD came down in a pillar of cloud and stood at the doorway of the tent, and He called Aaron and Miriam. When they had both come forward, He [God] said, 'Hear

---

13   Despite Greek and Roman cultures supporting polytheism, a hierarchy does exist in the gods, with Zeus (whose Roman name was Jupiter) at the top of the hierarchy. Zeus had a wife called Hera who became higher in the hierarchy than did other goddesses and quite likely other gods. However, it is generally agreed that Zeus was at a higher level in the hierarchy compared to Hera, just like his father Cronus was until Zeus took over. Thus, technically, one could replace "Zeus" with "God". The only problem with these cultures is that many people did not realize there was only one God and instead chose to idolize and worship many different gods. This polytheism led to problems later with people arguing their god was more powerful than was another, and each person acquired a different set of characteristics related to their chosen god, some of which did not appropriately promote the true principle of love.

The Ancient of Days 1794 Relief etching with watercolor; 23.3cm x 16.8cm, by William Blake (1757–1827). British Museum, London, Great Britain. (Credit: akg-images / Erich Lessing).

now My words: If there is a prophet among you, I, the Lord, shall make Myself known to him in a vision. I shall speak with him in a dream. Not so, with My servant Moses, He is faithful in all My household; With him I speak mouth-to-mouth [or face-to-face], even openly, and not in dark sayings, and he beholds the form of the Lord. Why then were you not afraid to speak against My servant, against Moses?'

So the anger of the Lord burned against them and He departed. But when the cloud had withdrawn from over the tent, behold, Miriam *was* leprous, as *white as* snow. As Aaron turned toward Miriam, behold, she *was* leprous.

Then Aaron said to Moses, 'Oh, my Lord, I beg you, do not account *this* sin to us, in which we have acted foolishly and in which we have sinned. Oh, do not let her be like one dead, whose flesh is half eaten away when he comes from his mother's womb!'

Moses cried out to the Lord, saying, 'O God, heal her, I pray!'

But the Lord said to Moses, 'If her father had but spit in her face, would she not bear her shame for seven days? Let her be shut up for seven days outside the camp, and afterward she may be received again.'

So Miriam was shut up outside the camp for seven days, and the people did not move on until Miriam was received again."

But Moses was not the only person to think this way. The Bible mentioned other people who had heard or seen God. None of them hinted at a feminine God. It appears, then that God had chosen to communicate to Moses and other humans in the Middle East in the masculine form. It is as if God felt being masculine was the only way to influence a male-dominated human society.

However, when broadening our information sources to look beyond the Bible, evidence supports attributing a feminine quality to the one true God.

In fact, Eastern mystics seem to support the idea of a balanced God. They acknowledge the dualistic property of God that must somehow transcend to become unity in order for the Universe and

257

Ezekiel's vision of God, painted in 1518 by Italian painter and architect Raffaello Sanzio (1483–1520) in the Palazzo Pitti in Florence, oil on wood.

ourselves to exist. Thus, if describing God as masculine, we must also acknowledge God's feminine side. On this idea, Jagassar Das, Kabir Satsang of Canada, said the following:

"God is neither male, nor female, nor neuter gender. God is neither good nor bad; He is beyond the dualities of the material world. God is the Power that 'Just Is'."[14]

If we need further quotes to support God as having a feminine side and with names suggesting this quality such as "She" and "The Great Mother", we know Hans Küng discussed this issue quite openly in *Does God Exist?*:

"The history of religion has shown that God can be called 'father' in a great variety of religions....But is it not striking that in the matriarchal cultures, in place of the Father God there is the 'Great Mother', out of whose fertile womb all things and beings emerged and into which they return?
...God then, is not [just] masculine....The feminine-maternal element must also be recognized in him."

So now, we must face the fact that God is neither solely feminine nor solely masculine, but both. We must, therefore, consider both genders when answering the question of who or what is God.

Where Can We Find God?

This issue of God is definitely proving itself to be quite difficult to pinpoint in a physical and precise form. Despite Moses in the Bible claiming to have seen God and describing this entity in the masculine form, the broader religious view, as shown in the prior section, is that God is neither feminine nor masculine, but rather a combination of both. And this duality is also, for some reason, pushing us to see God in both the opposite extremes and from a central balanced position for unity. How odd? So let us have another look at this God issue from another angle.

What about asking where God can be found? In other words, is there a specific location God tends to hang out (like some people do at

---

14   http://mysticson.blogspot.com.au/2007/07/gender-of-godby-james-bean.html.

a pub or club) more often than any other place? Because if so, perhaps we can train our eyes to observe this mysterious entity in great detail. Or it might reveal details of the true nature of God if we can localize this entity in place.

In ancient Greek mythology and Christianity, a general consensus exists among the religious leaders and many of their devoted followers and supporters. In their view, God is probably located somewhere in the heavens, but not necessarily anywhere specifically on Earth. Well, there are a couple of exceptions. As mentioned earlier in this chapter, in Christianity we have heard the story of Moses and his people having witnessed a large cloud in the sky under intelligent control, as if God resided in it. Then we have some people who have claimed that God may have appeared in the flesh through a charismatic leader named Jesus Christ more than 2,000 years ago. Leaving aside the earlier observation and the latter argument from biblical scholars, there does not appear to be any other location where God can be found, except somewhere above our heads.

On the other hand, if we look at the religious knowledge shared by Eastern mystics and even those supporters who have deeply thought about the question, we find a truly odd thing. No one can say God is only in one place.

For example, in Hinduism, religious leaders state that "God can be found in each one of us".

In other Eastern religions, this idea has been significantly broadened to say that the one true God is a mysterious entity located in all things and within each one of us. Meher Baba writes in his book *Life at Its Best* in support of this view:

> "God is everywhere, in everything. Most of all He is right within yourself."[15]

Baba's statement is similar to the one made by C. Birch and J. B. Cobb in 1981:

> "...God is not the world, and the world is not God. But God includes the world, and the world includes God....There is no world apart from God, and there is no God apart from

---

15   Vardey 1995, p.135.

some world.....God's life depends on there being some world to include."[16]

Perhaps the best way to understand this tricky concept is to imagine a hologram. When the hologram is broken up into many tiny parts, we can see some semblance of the whole picture in each part. However, we can never say with absolute certainty that any individual part is the complete picture; each part is only a representation of the whole. For us to see the true complete and holistic picture, we must combine all the parts[17].

Here is another example. In the lyrics of a song developed by a mystic, the person writes about God:

I am here and I am there and I show myself in different shapes.
And you may wonder what or who am I and you will not understand.
But in time the answer is given.
I am here and I am there and it is all the same,
Everywhere all the time, am I alone...[18]

And Thomas Merton writes the following in *No Man Is an Island*,

God approaches our minds by receding from them.
We can never fully know Him if we think of Him as an object of capture,
to be fenced in by the enclosure of our own ideas.
We know Him better after our minds have let Him go.
The Lord travels in all directions at once.
The Lord arrives from all directions at once.
Wherever we are, we find that He has just departed.
Wherever we go, we discover that He has just arrived before us...[19]

Not even trying to pinpoint God through time will help. On this, Hans Küng said the following in *Does God Exist?*:

---

16   Vardey 1995, p.806.
17   Again another reason why humans must unify world religions if there is any hope to better understand the true nature of God.
18   Vardey 1995, p.457.
19   Vardey 1995, pp.28-29.

"…he is not only God for me here and now, but God at the beginning, God from all eternity."[20]

Since God seemingly appears a bit all over the place and with no moment in time for us to definitely identify God as being present, we find ourselves in a rather difficult and unpalatable position. Does this mean that God does not exist because we are being led up the garden path by bizarre statements? Not according to the most knowledgeable and trusted religious leaders. They are certain God exists. Indeed, they claim that there is virtually no way of escaping God's presence in the Universe.

So what are we to make of these strange insights by religious leaders? And can we ever truly figure out where God can be found, and hence answer the question of what is God?

Is God a Paradox?

Plenty of evidence in religious leaders' teachings suggests that God is neither just this nor just that, but yet somehow includes both this and that.

For example, the idea that God is old seems to be questioned when some Christians claim God appeared in human form as Jesus more than 2,000 years ago. As God just so happened to be a young person, shouldn't God be called "the Young One"? Or is "the Old One" meant to be a metaphor for someone with wisdom and not necessarily tied to one's age?

What about the gender of God? As mentioned earlier in the chapter, we cannot say with certainty that God is either masculine or feminine. God could, in fact, be both.

As for saying whether God exists at a specific place or moment in time, we are told that God is everywhere and throughout all times. Yet, at the same time, God is within each one of us right at this moment in time.

Sounds pretty confusing.

Or maybe not. What if there is a point to this to-ing and fro-ing from one extreme to another? For a start, we know that God can be described as both young and old, feminine and masculine, omniscient

---

20   Vardey 1995, p.13.

and not omniscient, and so on. We could actually continue in this odd way if we so wish. For instance, God is either absolute experience or absolute knowledge. God has a name, and God does not have a name. God is here, and God is there.

Something really odd is happening here. We cannot help noticing a paradox developing in the views of God given to us by religious leaders.

Does this mean God is meant to be a paradox, and religious leaders are trying to explain this paradox in a slightly long-winded approach through the words they use?

Indeed, if we press for more information, we are often forced to consider this paradox as if God is meant to be a paradoxical entity. The religious thinkers present God as a paradox by saying things like, "God is one and God is many", "God is here and God is there" and "God is within us and God is everywhere". Such paradoxical statements are used because it seems this is best way to describe God by using the words available to us.

Consider three more paradoxical statements:

1. God is known, and God is not known.
Thomas Matus writes in a paradoxical manner about God when he says: "...of knowing God by unknowing..."

The idea of knowing God by unknowing is supported by J. Krishnamurti in his book *On God*: "God is something that cannot be talked about, that cannot be described, that cannot be put into words, because it must ever remain the unknown."

2. God is attainable, and God is not attainable.
Vaclav Havel wrote in *Letters to Olga* that "...we know of the unattainability of what we strive toward, and we know that we cannot help but strive toward it..."

3. Hatred of God brings love from God.
John Cornwell writes another paradoxical statement in *Powers of Darkness, Powers of Light* another paradoxical statement: "Hatred of God may bring the soul to God."

Jacob's Ladder, or Jacob's Dream, c.1799–1807 by William Blake (1757–1827)
© *The Trustees of the British Museum.*

That statement might explain why some people say to suffer is a great gift of God, for it teaches us how to love. God hurts only to heal. In the words of Mother Teresa of Calcutta, as told to us by Edward Le Joly,

suffering is a great gift of God;
those who accept it willingly,
those who love deeply,
those who offer themselves
know its value.

Is God meant to be a paradox?

What is the point of God being a paradox if we don't know whether God is this or that? What is this supposedly teaching us?"

No one knows for sure. To the more experienced and/or knowledgeable religious leaders, this paradoxical nature of God is probably just another way of saying that God is an "absolute truth" (i.e., the unity or oneness we are seeking to understand and experience) that satisfies two oppositely described and experienced half-truths of life and the Universe (i.e., the the parts of the hologram that we spoke of earlier). It is claimed that our purpose is to approach this "absolute truth" (we shall call it God for now), but never can we actually attain it —and certainly not in an immediate sense. Whether or not we are aware of it, we must, as part of our life's work, experience and learn both opposite half-truths of life and the Universe in a switching or cyclic manner over time as a way of approaching this absolute truth. But never can we simultaneously see both half-truths in order for us to truly know (and hence fully experience) the absolute truth (i.e., God itself).

For example, the Christian artist William Blake (1757–1827) illustrated the cyclical concept of life, depicting how we approach some ultimate goal and with it the true understanding of God. In his painting *Jacob's Ladder*, we see how Jacob becomes spiritually enlightened during his sleep, helping him to see the meaning of life. Notice how the artist uses a spiraling stairway heading towards a bright light described as our destiny.

Is this light at the top supposed to represent God? And is the spiraling staircase a kind of cyclic switching between two half-truths in order to approach an understanding of God?

To better appreciate the concept, take a glance at the picture below,

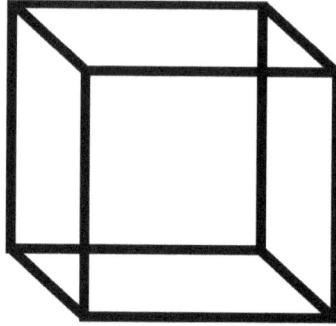

You will find within this picture an image of a cube from two perspectives: one cube viewed from above and another viewed from below.

Did you have to tell your brain to switch between the two images in order to properly recognize them? Don't worry, it is perfectly normal. This switching effect in your mind, no matter how quickly you try to do it, is actually quite natural. It also forces you to realize that you can never simultaneously see and recognize the two boxes from two different perspectives simultaneously in your mind. Yes, the two images are clearly sitting there on the page, but to recognize this fact, you had to tell your brain to consciously first create one image of a box from one perspective before you could create the other box from the other perspective.

The same dynamic is true of life and the Universe. We need to switch constantly between two half-truths in order for us to see the ultimate Truth if given enough time.

In the case of God, the same switching between two opposite extreme half-truths seems to be a necessary part of understanding this entity as a whole.

The Link Between God and light

With God looking like a paradoxical entity, there is an opportunity for science to determine whether the religious leaders are right. All we have to do now is find an equivalent paradoxical entity in the world of physics that could support the religious concept of God. Should this entity exist for the scientists, a link could be established between religion and science through this God concept, and so give greater credence to the concept by religious leaders, and perhaps even provide more insights and a better understanding of this mysterious entity. As author Paul Collins said:

"I mean we can't define God, the best that we can do is gain some insights into who or what God might be."[21]

Fortunately, one insight we already have is the fact that God is a paradox. In which case, in the words of the Danish physicist Niels Bohr (1885–1962):

"How wonderful that we have met with a paradox. Now we have some hope of making progress."[22]

Progress indeed. Because it is now possible to find in the unified field equations a paradoxical entity ubiquitous through the Universe. Remember that thing we called light? You know. The electromagnetic energy we have been talking about throughout this book and forms the very essence of Einstein's Unified Field Theory. Well, there is something special about it. You see, radiation (or in its most common form called light) is also described by many scientists as a paradoxical entity in its own right. For example, we know light is both a wave and a particle. In fact, if we replace the word "God" with "light", we can better understand the many paradoxical statements religious leaders have come up with.

For example, according to Einstein's General Theory of Relativity, the mass-energy of space-time exists everywhere we look. From the furthest reaches of the visible universe and beyond to the atoms composing us and the chemical bonds holding the atoms together, we

---

21   ABC Radio National, January 30, 2011. Transcript from
      http://www.abc.net.au/rn/encounter/stories/2011/3120525.htm.
22   Moore 1966, p.196.

know all mass-energy is the same. In the Unified Field Theory, this mysterious mass-energy is light. Since we are all made of the same stuff (i.e., electromagnetic energy), it is technically correct to say that light exists within us and in all things around us. The concept of light thus fits in well with the religious idea that God is inside of us and in all things around us, but never is all of this energy located in just one thing.

We also find that light has the ability to influence other matter and increase the likelihood of colliding and staying together in what was previously called gravity and universal gravitation. In the Unified Field Theory, however, it should be seen as a kind of "pushing" force of radiation rather than the "pulling" force of a mysterious gravitational field mentioned by traditional 20th century scientists. So in a sense, light brings together and guides all things towards a sense of greater order and complexity. To religious leaders, God is said to be doing a similar thing with everyone and everything around us; it is trying to push us (or guide us) toward some ultimate goal, as religious leaders would say.

Interesting idea.

But apart from these interesting religious sayings about God and finding equivalent examples in physics to support the sayings using the concept of light, the paradoxical nature of light is the most intriguing aspect here. Indeed, it is a paradoxical game that is being played in front of our eyes when we go to the extremes of human observation at the grandest scales of the visible universe and at the smallest scales of quantum particles.

In particular, we have seen how the redshifting effect of light from the furthest galaxies (see Chapter 6) has created a paradox in the size and age of the universe. Either the visible universe is finite and expanding or we live in an infinite Universe depends on how we interpret the redshifting effect. You can explain the redshifting as a natural energy loss as light travels through the radiation-filled space, or the Doppler effect of objects receding from us. Whichever of these two explanations we choose, it seems the light is telling us something: the further we look into the visible universe, the more paradoxical the universe seems to get.

Likewise, in quantum theory, the quantum particles exhibit a paradoxical behavior of either a "particle" or a "wave" when one slit or two slits are opened in the modified Thomas Young experiment. We suspect this result is due to light from the environment influencing the quantum particles to behave paradoxically. Even so, we still see a behavior in particles that reminds us most clearly of the nature of God. In fact, it is funny to hear the unresolved debate between Einstein and those of his colleagues supporting quantum theory about whether God is somehow influencing the particles, or the particles themselves are acting as God in knowing where to go when the conditions are set up in a certain way. God is certainly helping to create a paradoxical result in quantum theory.

Or should we say that it is light doing all the work?

And now, when we do focus more closely on light, we discover that light itself is indeed a paradoxical entity. Interesting. Why should light (and the way it influences other matter at the extremes in observations) behave paradoxically as if it is God?

Could light be God?

Or perhaps light is like a broken piece of a hologram, and for us to see "the mind of God" we must combine all of the pieces in the visible universe (and possibly beyond) to reach the final answer. If so, maybe God is actually the Universe (or the ultimate hologram representing true reality)?[23]

Whatever the situation, it seems like the religious leaders may have been right all along to say God is both the unity of all things and is paradoxical in nature due to its dualistic properties. In science, it is finally looking like light is also what unifies everything when creating the reality we see call the Universe while revealing its paradoxical behavior (and with it the importance of balance between opposite extremes) through the oscillations in the mass-energy density moving from one extreme to the next and back again.

Knowing that light and God both behave like paradoxical entities bringing unity in the Universe, and forming a sense of perfect balance to everything, does this mean God and light are one and the same thing?

---

23   Does this mean understanding the "mind of God" is all about understanding what the Universe has done in the past, what it is doing now, and what it is going to do in the future, as well as seeing what the entire Universe looks like? Is this practical? How big is the Universe to know whether it is doing anything? If so, have we given ourselves enough time and traveled far enough into space and time to gather all the evidence and interpreted that evidence correctly to know what is happening in the Universe?

Certainly we cannot have light as one God and God itself as another separate entity, both existing at the same time and in the same Universe. As we are told by religious leaders, there can be only one God. So either God is light, or light is mimicking the characteristics of God, like a tiny piece of a hologram. If it is the latter, this must imply God is something else more mysterious, and that we need light in space and in matter to be combined to see the entire Universe, and with it, the true nature of God.[24]

Is there any further evidence to support the link between light and God?

Interestingly, we find in the Bible no less than 272 examples of "light" being used to represent God in the New Testament. Here are some examples:

> "I am the light of the world." (John 8:12)
> "In him was life, and that life was the light of men." (John 1:4)
> "Ye are the light of the world." (Matthew 5:14)

And in the Old Testament, of course we have the following example:

> "Let there be light" (Genesis 1:3)

Here is another example taken from the Old Testament:

> "Live as children of light (for the fruit of the light consists in all goodness, righteousness, and truth) and find out what pleases the Lord. Have nothing to do with the fruitless deeds of darkness, but rather expose them."[25]

It seems we have reached a controversial pinnacle in this research on the Unified Field Theory. Have we found a way for science to understand God by looking at the electromagnetic phenomenon we call light, thanks to its paradoxical property?

---

24   Another analogy involves observing the multitude of dots forming an Australian Aboriginal painting. The closer you approach the painting, the more you see the individual dots in detail, but the less obvious the big picture becomes. One needs to step back to observe all the dots in order to see the whole picture.

25   Ephesians 5:8–11.

The View of Einstein on the God Issue

So what does this mean for Einstein and the God issue? Do he give his support to the concept of God? What did he find out about God after completing his Unified Field Theory? The answer to these questions depends on how you define God. In a letter to a friend in 1954, Einstein said the following:

> "...The word God is for me nothing more than the expression and product of human weaknesses, the Bible a collection of honorable, but still primitive legends which are nevertheless pretty childish. No interpretation, no matter how subtle, can (for me) change this..."

What Einstein truly disliked was the view of "God" as some anthropomorphic entity who could be sitting in the sky watching over and influencing human events[26]. Given our understanding of the concept of the one and only true God of the Universe present everywhere and not confined to any one region of space and time, this view is perfectly reasonable. Whatever happened in biblical times to influence certain people to think in a more simplistic way given their claims of encountering God in the Middle East suggests that another "god" may have been involved.

Without proof of the existence of extraterrestrial life and the technology needed to reach our planet from distant worlds outside our solar system, however, the scientific side of Einstein must have taken over and told him this occurrence was unlikely. As a result, he was an agnostic on the issue and, by the age of 12, had already made up his mind regarding a majority of the stories he read in the Bible. As Einstein said:

> "...I came—though the child of entirely irreligious (Jewish) parents—to a deep religiousness, which, however, reached an

---

26   Einstein's view of a "personal God" came at a time when lille was published in newspapers of mysterious symmetrical, metallic, and glowing flying objects with distinct electromagnetic effects appearing together with the appearance of humanoid occupants. Whether these are examples of alien civilizations visiting the Earth does warrant further research. But one thing is certain: a sufficiently advanced alien civilization could be seen by scientists as virtually indistinguishable from God in terms of technological capabilities. And if so, there is no reason that God could not have been defined in the past as one of these civilizations capable of influencing human events. Just such a scenario may have taken place at least on a couple of occasions in biblical times, and perhaps after 1947. It is, therefore, important for us, as scientists, to always keep an open-mind.

abrupt end at the age of twelve. Through the reading of popular scientific books I soon reached the conviction that much in the stories of the Bible could not be true. The consequence was a positively fanatic orgy of freethinking coupled with the impression that youth is intentionally being deceived by the state through lies; it was a crushing impression. Mistrust of every kind of authority grew out of this experience, a skeptical attitude toward the convictions that were alive in any specific social environment—an attitude that has never again left me, even though, later on, it has been tempered by a better insight into the causal connections. It is quite clear to me that the religious paradise of youth, which was thus lost, was a first attempt to free myself from the chains of the 'merely personal', from an existence dominated by wishes, hopes, and primitive feelings. Out yonder there was this huge world, which exists independently of us human beings and which stands before us like a great, eternal riddle, at least partially accessible to our inspection and thinking. The contemplation of this world beckoned as a liberation, and I soon noticed that many a man whom I had learned to esteem and to admire had found inner freedom and security in its pursuit. The mental grasp of this extra-personal world within the frame of our capabilities presented itself to my mind, half consciously, half unconsciously, as a supreme goal. Similarly motivated men of the present and of the past, as well as the insights they had achieved, were the friends who could not be lost. The road to this paradise was not as comfortable and alluring as the road to the religious paradise; but it has shown itself reliable, and I have never regretted having chosen it."[27]

In a more direct statement regarding Einstein's view on a "personal God", he said the following:

"I cannot conceive of a personal God who would directly influence the actions of individuals, or would directly sit in judgment on creatures of his own creation."

---

27   Einstein 1979, p.3.

However, after working on the Unified Field Theory, Einstein came to believe in what he called "Spinoza's God", which is basically a form of pantheism[28]. This belief appears in a letter Einstein wrote to Rabbi Herbert S. Goldstein of the Institutional Synagogue, New York, as a direct response to Goldstein's five-word cablegram, "Do you believe in God?". The letter, dated April 24, 1929, contains the following interesting quote:

> "I believe in Spinoza's God, who reveals himself in the orderly harmony of all that exists, not in a God who concerns himself with the fate and the doings of mankind."[29]

In other words, Einstein must have discovered something in his Unified Field Theory to make him realize that a ubiquitous substance permeates all of space and in all solid matter, and that it provides the unifying force for connecting everything together and showing "the orderly harmony of all that exists". Thus, if this substance is light (and just so happens to be a paradoxical entity as we see for God in religion), it would not be too far off the mark to say that Einstein probably did believe in God. It all depends on how we define God in a scientific sense, not necessarily in the way certain Western religious people have created in their own minds.

Hence this would probably explain why Einstein said:

---

28   Pantheism is essentially monism, or the unity of all things. But whether this unity reveals some kind of a preordained plan for everything depends on how religious or scientific your views. In its fundamental concept, monism was made prominent by the 17th-century philosopher and staunch religious thinker Baruch Spinoza, who published his book *Ethics* as a direct response to Rene Descartes' "dualist theory". But whether pantheism can be interpreted in a religious and non-religious way depends on the individual. From a religious position, pantheism can be viewed as an all-encompassing and infinite God existent inside and interconnecting all things that, if seen in its totality, should reveal an "intelligent plan". From a non-religious position (probably how Einstein would have preferred), God is nothing more than the Universe itself, which may or may not have a grand plan, and some kind of a universal substance exists that makes up and interconnects all things, whatever that might be. So when scientists talk about seeing the "mind of God" in their work as if trying to establish a possible intelligent plan in the Universe that could be attributed to a mysterious entity called God, in pantheism, it is not a ray of light in isolation that is God (even though it will behave like God in many respects). That statement is just like saying a living cell existing in isolation from all other things is not God, but rather it is the totality of everything that we see (i.e., all interconnected living cells and living creatures). Only then would we know the truth. Until we see the entire Universe, all scientists can do is write a simplistic representation of God in an equation using the oscillations of radiation to help show this paradoxical entity. Beyond that, each person must decide what to believe is the truth: is there a grand plan for the Universe and for all of us, or none at all?

29   *The New York Times*, April 25, 1929, p. 60, col. 4. Ronald W. Clark (pp. 413–414)

"The main source of the present-day conflicts between the spheres of religion and of science lies in the concept of a personal God."[30]

Fortunately, it need not have to be this way. The Unified Field Theory has shown us a way to define God differently and still be consistent with the statements of religious leaders. If it turns out this is an acceptable approach, even by scientific standards, the potential is there to finally end the controversy and resolve all "present-day conflicts" on the matter, and so unify science and religion on certain concepts (not just on God, but on other aspects made by the mystics as shown in Appendix C). Of course, whether it actually does end up being this way will be totally up to everyone to decide if this is the way forward for humankind.

---

30   Bernstein 1991, p.20.

CHAPTER 9

Conclusion

You are a part of the Universe, no less than the stars and trees,
and you have a right to be here. And whether it is clear to you or
not, no doubt the Universe is unfolding as it should...
                                                    —Desiderata

HAS THE holy grail of physics been found thanks to Einstein's work on the Unified Field Theory and the discovery of a ubiquitous energy that seems to control everything in the Universe?

It may take some time for scientists to fully realize the implications of this theory even after reading this book. Nevertheless, the Unified Field Theory's ability to explain some of the biggest mysteries of science, such as gravity, from an entirely new perspective cannot be underestimated. All that is needed now is to test the ideas through experiments and to apply some simplified mathematics to see if the magnitude of the force of radiation would correspond to the real world.

Does this mean that Einstein succeeded in his ambitious attempt to unify physics using the all-pervasive field of nature known as *light*? And did he ultimately see the "mind of God" through his own work?

Professor Hawking thinks Einstein may have been unsuccessful mainly because little was known about both the weak and strong nuclear forces during his time. As Hawking writes in his book *A Brief History of Time*:

> "Einstein spent most of his later years unsuccessfully searching for a unified theory, but the time was not ripe: there were partial theories for gravity and the electromagnetic force, but very little was known about the nuclear forces. Moreover, Einstein refused to believe in the reality of quantum mechanics, despite the important role he had played in its development."

Then again, it is possible that scientists themselves have only given themselves a fleeting consideration to the reality of the Unified Field Theory and how it relates to atomic physics. To give an example of how the Unified Field Theory will be important to atomic physics, let us use the concept of light to look at the mystery of how protons can stay together in the nuclei of atoms.

The first undeniable fact that scientists have established is that protons are positively-charged particles. The next fact comes from the laws of electromagnetism—namely, like charges repel. In that case, protons should do the same, except inside a nucleus, they do not. How is this possible?

On closer examination, scientists have noticed another subatomic particle called the neutron existing in the presence of two or more protons in the nucleus. With no other particle to consider, it seems rather obvious that this neutron must play an important role in keeping the protons together. The question is, *how?* Neutrons have no apparent electric charge. In fact, by definition, scientists have accepted the "uncharged" view, based on the results gathered by their instruments. But an uncharged particle does not attract the protons. Yet the neutrons must exert a mysterious force on the protons to keep the nucleus together. The best scientists can do for now is to conceptualize a unique force of nature called the *strong nuclear force* to get around this problem.

Then again, the strong nuclear force may have a simple classical electromagnetic explanation. If one exists, it would show that the Unified Field Theory does have validity in the world of atomic physics.

Let us see if this is true.

After further analysis using a particle accelerator, it has been confirmed that the neutron is fundamentally composed of an electron and a proton, with a little extra energy known as an anti-neutrino acting as a kind of electromagnetic spring, or in biology, a kind of disc separating two vertebrate bones to help prevent the electron merge with the proton. This is the simplest picture we are presented with. It is either that, or we could always go for the complicated picture being used nowadays by the physicists, which is as follows:

> "Neutrons and protons are classified as hadrons, subatomic particles that are subject to the strong force and as baryons since they are composed of three quarks. The neutron is a composite particle made of two down quarks with charge ⅓ e and one up quark with charge +⅔ e. Since the neutron has no net electric charge, it is not affected by electric forces, but the neutron does have a slight distribution of electric charge within it. This results in non-zero magnetic moment (dipole moment) of the neutron. Therefore the neutron interacts also via electromagnetic interaction, but much weaker than the proton."[1]

It is probably better that we stick to the simpler picture.

Now, when these particles are combined together to form this thing called a "neutron", a natural assumption would be to say that the neutron is uncharged, right? As they say, an electron and a proton are both electrically charged, and the electron has the exact opposite charge to a proton. Combining the two should, therefore, cancel out the charge.

Scientists could be forgiven for thinking in this way; after all, this is what they see with their instruments. If the instruments do not show a charge, why would scientists consider any other possibility? Just assume

---

1   https://www.nuclear-power.net/nuclear-power/reactor-physics/atomic-nuclear-physics/fundamental-particles/neutron/structure-neutron/

there is no charge at any instant in time, and life would be simple. Well, not quite. Scientists are not finding it simple because they are having a hard time explaining what this strong nuclear force actually is and how to combine it in a unified way with the other forces of nature.

So, what do we do now?

Let us consider this idea: What if the electron and proton remain separate entities and are performing some kind of electromagnetic dance inside the neutron, so that the charge of the neutron rapidly swings from negative to positive? Now, here is an interesting thought. Because if we think carefully about this new "picture" of the neutron, how can scientists determine for sure and with precision from their instruments whether the neutron is truly uncharged? The problem with such instruments is that they are very large, especially the electrode for measuring the electric charge, and the mechanism of the instruments to sample the electric charge is not fast enough to detect the change. Thus, what we have here are oversized, sluggish, and cumbersome instruments providing an average result, which gives the impression that the neutron has no charge. Yet in reality, the neutron is quite likely constantly changing from negative to positive so quickly that we cannot detect this change.

So, as the neutron swings into the negative charge (or at least over one end of it) and another proton comes into range, there is a possibility that the electron hidden inside the neutron can momentarily pop out to join the other proton. A new neutron is formed, and the remnants of the old neutron suddenly become a proton. It is this transition of the electron that could momentarily help to electromagnetically attract another proton just prior to creating a new neutron. If scientists were not privy to this new "picture" of the neutron, then the next question to leave them scratching their heads is, "Why would an electron come out of a neutron if it is so closely bound to the proton, or possibly merged to form the neutron?" Without a clear mechanism to measure what could be happening, scientists would be forced to describe this transformation of a neutron into a proton as a kind of beta decay due to a *weak nuclear force*. Yet another exotic force of nature is created to keep the scientists busy. However, again the movement of the electron from the neutron to a proton is probably nothing more than a natural electromagnetic force between protons and electrons.

Now, what of this strong force holding the protons together? What exactly is this?

Looking more closely at the nucleus, we make one interesting observation: the process of an electron popping out of a neutron and joining up with a proton is repeated many times at an extremely high rate of speed at very close range. Because this process repeats itself so quickly and the particles are so small, extremely lightweight, and move over such short distances, the nucleus is probably unable to break apart. The constant electromagnetic forces appear and disappear so rapidly that they help to keep the electrons and protons together. In a sense, the electrons (or we may say the electromagnetic field "pushing" unlike charges together) may be helping to "glue" the protons together.

For an analogy, imagine a circus clown trying to keep three large balls floating in the air. The clown gives the impression that the balls are floating because he uses his hands to temporarily squeeze two of the balls together with the third ball being held in the middle. Suddenly, he moves his hands away and, for a moment, it seems like the balls are floating in the air, during which time it is feasible to quickly pick up one or two balls in one or both hands and move them around to a different position. Of course, the remaining ball still "hanging" in the air will not stay in mid-air forever. Gravity will start to act on the ball to bring it down to the ground. But before it does move, the two other balls are brought back together again to hold the third ball in position. The process can be repeated as many times as the clown wishes, as long as he does it quickly enough to ensure that none of the balls move apart or fall to the ground.

At the same time, it doesn't matter how much force is applied to keep the balls together in the air, even a small amount of force is sufficient. The aim here is to move things around very quickly to ensure the balls don't fall down.

In this case, a similar trick likely happens in the nucleus. The protons are naturally jostling around within the nucleus, but the forces of the electromagnetic field between neutrons (or more precisely, electrons) and protons are probably helping to prevent the protons from flying off into space.

Still, the nucleus has one more interesting trick up its proverbial atomic sleeve. Scientists say that the nucleus is spinning on its axis. Since the nucleus is positively-charged, the laws of electromagnetism tell us that an accelerating/spinning charge must emit *synchrotron radiation*. The presence of this radiation helps to perform three important tasks:

1. To create a natural electromagnetic recoiling force for pushing the protons and neutrons into a tightly packed structure, making it easier for the electrons in the neutrons to do their job of keeping the protons together;
2. To create concentric "gravitational" standing waves around the nucleus of decreasing strength with increasing distance to hold additional electrons in specific orbits around the nuclei, and to help neutralize the positive charge of the nuclei as well;
3. To act as an electromagnetic barrier against the ravages of universal background and man-made radiation from outside that can cause *radioactive decay*[2].

With these explanations out of the way, it now appears remarkably like the weak and strong nuclear forces are purely electromagnetic in character and certainly nothing exotic. The electromagnetic fields appear to be doing all the work of keeping the protons together simply by the humble actions of a proton quickly snatching an electron from a neutron, thereby allowing the attractive forces of the electromagnetic fields to appear for a fleeting moment.

How interesting.

However, as with any new explanation, it will need to be tested to see if this is true. Until then, it is clear that scientists have a considerable amount of work to do in this field to be absolutely certain.

---

2   A fast enough spin rate of the nucleus should create strong enough synchrotron radiation to prevent natural (or artificial) radioactive decay of the nucleus. Radioactive decay is caused by sufficiently high-frequency radiation from the environment splitting the nucleus apart, and the spin rate of the nucleus is slow (usually the case for heavy elements). Consequently, natural radiation in the environment will be able to penetrate the nucleus to create a natural radioactive decay of these heavy elements over time. However, change the mass-energy density of this universal background radiation, and you can increase the amount of high-frequency radiation, causing high-spinning nuclear elements previously considered stable such as carbon, oxygen and nitrogen to become radioactive in this environment. Similarly, in a low-radiation energy density environment, it might be possible to reduce the natural radioactive decay rate of a slower spinning neutron-rich and unstable atomic nucleus. In other words, lowering the temperature of heavy elements could lower the rate of radioactive decay.

Despite this, we can already see how Einstein's Unified Field Theory will make a significant contribution to atomic physics.

Although this book reveals the need for further work on the Unified Field Theory and its impact on all areas of science, greater insights into many of the more significant problems of physics, chemistry, biology, and astronomy are already apparent. Most importantly, we see that the solution to such problems does not require the introduction of exotic new particles, wormholes or singularities in space, traveling back in time, or other amazing suggestions made by the mathematicians who are solving Einstein's gravitational field equations. Instead, we can now find classical Newtonian explanations for various problems, thanks to the presence of the gravitational field in the electromagnetic field and finding new electromagnetic pictures of how solid matter might behave.

So where do we stand in terms of Einstein's final great theory? In other words, was he successful? This depends on the problem you are trying to solve.

We are now on the cusp of finding new explanations in the world of science simply by focusing on the laws of electromagnetism as this will give physicists the best chance of unifying all of physics. However, as far as the "mind of God" is concerned, probably not. The concept of God does have a scientific equivalent in physics if one relies on the concept of light. But whether we can understand the "mind of God" requires combining all electromagnetic energy in the Universe, and that means we must know everything about this Universe and what it might be doing. The reality, as we have seen in Chapter 6, is that no one really knows what the Universe is doing. So how can we know "the mind of God"?

What we do know is that Einstein did understand the concept of God from what he called Spinoza's God. Here, he appeared to have acknowledged a ubiquitous energy that permeates the Universe and creates all solid matter and creates all the laws of physics. Following Einstein's work on the Unified Field Theory, we can see how light is essentially this type of substance. However, in terms of understanding the "mind of God" to help him see things such as what the visible universe (let alone the larger, unseen Universe) is doing, where we are going, and why we are here, it seems our beloved Einstein may have run out of time and faced insurmountable physical hurdles to reach this ultimate level.

Perhaps we should not be surprised by this finding. After all, Einstein was just a man, not God. There can only be one God—the thing we are striving toward in our work. Einstein is definitely a god given his influence in physics, but he is not anything more than this. Furthermore, religious leaders (also considered gods in their own right given their influence on the followers of the religion) have spent many thousands of years studying the problem of God and, although they may agree on the existence of God, they are still unable to determine precisely what God *is*, other than realizing it is a paradoxical entity. Now, in the 21st century, scientists have a new way to understand God by looking at light and all its properties. The question is, how long will it take for scientists to finally solve this God problem with the help of light?

In other words, is light God? Or is light like God (like a piece of a hologram), and we must combine all electromagnetic energy to see the Universe as a whole? If it is the latter, given how well scientists have interpreted the redshifting effect in the light from distant galaxies and making the claim that the universe must be expanding (see Chapter 6), we should not be surprised if solving the "mind of God" will take a very long time.

As for understanding what light is in its most fundamental sense, we also have a long way to go. For science, all we can say now is that light is just another piece of ordinary matter influencing other matter, and that light does behave in a paradoxical manner, just like God does for religion. Sure, light is an electromagnetic field, but what exactly is this field? Without resorting to mathematical concepts, no one really knows. It is just a mathematical entity that exists in the real world and with no obvious way to understand it in reality, other than to accept its solid matter behavior when it collides with other matter. Apart from its ability to exert a force on solid matter, we are far from knowing the truth about the nature of light.

The same is true for God.

Despite Einstein probably failing to see the "mind of God", we can still say with reasonable confidence that Einstein did successfully achieve one very important thing for all of us. Not only did he help us to unify physics under the umbrella of electromagnetism, making this the fundamental physical law of the Universe for explaining all phenomena from the smallest to the largest, but thanks to his Unified Field Theory, he may well have bridged the gap between religion and science.

Would this be enough to work toward unification by all religious leaders, just as scientists are doing right now?

Until then, it should be clear that when scientific and religious knowledge is simplified and unified, we will be able to see what it means for religious leaders to talk about God in religion and of light for scientific leaders. Then all scientific and religious leaders will at last work together on this common quest of understanding light, God, and the Universe as we finally forge an amazing future with a hope that will shine for many generations to come, especially for those who are alive and remain curious.

Indeed, for the first time in the more than 2,500 years since science split from religion in ancient Greece, a new era of cooperation could begin in the 21st century—and possibly, with it, a new world order for humankind.

Appendices

APPENDIX A

A Summary of Relativistic Mechanics

CONSIDER a stationary object of a certain mass, and all independent reference frames have coordinate axes parallel to and in the same direction as one another.

To an outside observer, he will measure the mass of that object as m, and any event that occurs on the object is positioned at a point (x_1, y_1, z_1, t_1) in his reference frame. To someone else who is sitting inside the object, he measures the mass to be m_o, and the position of that same event is (x_o, y_o, z_o, t_o) in her own reference frame. Since neither observer is moving with respect to each other, then the values of m_o, x_o, y_o, z_o, t_o equal exactly those of the outside observer.

Now if the object starts to move in the x_1 (or x_o) direction and manages to attain a velocity v, the position of any event on the object can still be calculated in either reference frame. However, neither observer in his or her own reference frame will agree on the position or time of that event. If the person inside the object, moving at velocity v, measures the position of the event as (x_o, y_o, z_o, t_o), the outside observer will measure the position and time of that event as follows:

$$x_1 = \frac{x_o - v.t}{\sqrt{1 - \frac{v^2}{c^2}}} \ , \ y_1 = y_o \ , \ z_1 = z_o \ , \ t_1 = \frac{t_o - \left(\frac{v}{c^2}\right)x}{\sqrt{1 - \frac{v^2}{c^2}}}$$

To the person traveling with the object, the mass of the object is m_o. However, the outside observer will measure the mass of the moving object as:

$$m_1 = \frac{m_o}{\sqrt{1 - \frac{v^2}{c^2}}}$$

Measuring momentum in different reference frames is again done in the same way: to the moving observer, the momentum of the object with respect to her is zero, and with respect to certain other objects in the Universe, momentum is equal to $p_o = m_o.v$. However, the outside observer will measure the momentum of the moving object as:

$$p_1 = \frac{m_o v}{\sqrt{1 - \frac{v^2}{c^2}}}$$

At the same time, all other objects in the Universe may (or may not) have zero momentum depending on which object the stationary observer is referring to in his calculations.

Suppose the person inside the object, still moving with velocity v, decides to throw a ball forward with velocity w with respect to herself in the x_o direction. To the outside observer, the velocity of the ball will be:

$$u = \frac{v + w}{1 + \dfrac{v.w}{c^2}}$$

If the moving observer switched on the engines of the moving object, a force is exerted on the object (and eventually the observer). The magnitude of the force may be $F_1 = m_o.a$ to the moving observer, but the outside observer looking in will not accept this fact because he calculates the force on the object to be:

$$F_1 = \frac{m_o a}{\left(1 - \dfrac{v^2}{c^2}\right)^{3/2}}$$

In the throwing-the-ball example, suppose the ball had a mass of m_o and moved with velocity w with respect to the moving observer. Its kinetic energy, K_o, according to the moving observer, is:

$$K_o = \frac{1}{2}m_o.w^2$$

To the outside observer, the kinetic energy of the ball with respect to him is:

$$K_1 = \frac{m_o c^2}{\sqrt{1 - \dfrac{u^2}{c^2}}} - m_o c^2 \quad \text{where } u = \frac{v + w}{1 + \dfrac{v.w}{c^2}}$$

The term on the left is called the total energy of the body, while $m_o c^2$ is called the body's rest energy. Total energy is usually given the symbol E_1 and can be related to momentum, p_1, of the body in the following way:

$$p_1 = \frac{m_o u}{\sqrt{1 - \dfrac{u^2}{c^2}}}$$

$$\frac{p_1}{m_o} = \frac{u}{\sqrt{1 - \dfrac{u^2}{c^2}}}$$

$$\frac{p_1}{m_o c} = \frac{u/c}{\sqrt{1 - \dfrac{u^2}{c^2}}}$$

$$\left(\frac{p_1}{m_o c}\right)^2 = \frac{u^2/c^2}{1 - \dfrac{u^2}{c^2}}$$

$$= \frac{1}{1 - \dfrac{u^2}{c^2}} - \frac{1}{1 - \dfrac{u^2}{c^2}} + \frac{u^2/c^2}{1 - \dfrac{u^2}{c^2}}$$

$$= \frac{1}{1 - \dfrac{u^2}{c^2}} - \left(\frac{1 - \dfrac{u^2}{c^2}}{1 - \dfrac{u^2}{c^2}}\right)$$

$$= \frac{1}{1 - \dfrac{u^2}{c^2}} - 1$$

But $E_1 = \dfrac{m_o c^2}{\sqrt{1 - \dfrac{u^2}{c^2}}}$ or $\left(\dfrac{E_1}{m_o c^2}\right)^2 = \dfrac{1}{1 - \dfrac{u^2}{c^2}}$

Therefore, $\left(\dfrac{p_1}{m_o c}\right)^2 = \left(\dfrac{E_1}{m_o c^2}\right)^2 - 1$

Simplifying this to the form $A^2 = B^2 + C^2$, we finally obtain:

$$E_1^2 = (p_1.c)^2 + (m_o c^2)^2$$

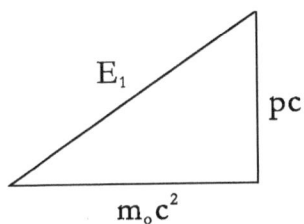

From here, the famous $E_1 = m_o c^2$ equation can be derived by setting momentum to zero in order to calculate the total energy of a non-moving object of mass m_o.

APPENDIX B

A Summary of Electromagnetism

AFTER EXPERIMENTS on electricity and magnetism were performed by Michael Faraday (1791–1867), and with contributions by Heinrich Hertz (1857–1894), James Clerk Maxwell (1831–1879) made a theoretical analysis of the electromagnetic knowledge and combined all the results into four of the most generalized field equations called *Maxwell's equations*. The equations relate charges and electric currents to two vector fields, **E** and **B**, representing the electric and magnetic fields, respectively.

The equations tell us the following:

1. An electric field **E** can exist without the presence of a magnetic field **B** if the charge is static.

2. A magnetic field **B** can exist without an electric field **E** if the current generating the magnetic field is constant.

3. If either the electric field **E** or the magnetic field **B** is not static, then one field cannot exist without the other and vice versa.
4. A magnetic field cannot be produced by a monopole (i.e., just a South or North pole of a magnet, for instance).

Maxwell's Equations

Maxwell's equations are:

$$\nabla \times \mathbf{H} = \mathbf{J} + \frac{\partial \mathbf{D}}{\partial t} \quad \textit{(Ampere's Law)}$$

$$\nabla \times \mathbf{E} = -\frac{\partial \mathbf{B}}{\partial t} \quad \textit{(Faraday's Law)}$$

$$\nabla.\mathbf{D} = \rho \quad \textit{(Gauss' Law for Electric Field)}$$

$$\nabla.\mathbf{B} = 0 \quad \textit{(Gauss' Law for Magnetic Field)}$$

These equations are written in the point form. In the integral form, these equations would be re-written as:

$$\oint \mathbf{H}.d\mathbf{l} = \int_S \left(\mathbf{J} + \frac{\partial \mathbf{D}}{\partial t}\right).d\mathbf{S}$$

$$\oint \mathbf{E}.d\mathbf{l} = \int_S \left(-\frac{\partial \mathbf{B}}{\partial t}\right).d\mathbf{S}$$

$$\oint_S \mathbf{D}.d\mathbf{S} = \int_v \rho\, dv$$

$$\oint_S \mathbf{B}.d\mathbf{S} = 0$$

Note that in a vacuum, $\mathbf{D}=\varepsilon_o\mathbf{E}$ and $\mathbf{B}=\mu_o\mathbf{E}$. And if the vacuum contains no flow of charges to create an electric current that might induce an additional magnetic field, $\mathbf{J} = 0$.

The symbols are defined as follows:

B	magnetic field (T or V.s/m^2)
J	current density (A/m^2) equal to $\mathbf{J} = \sigma\mathbf{E}$.
E	electric field (V/m)
D	electric displacement (C/m^2)
H	magnetic field intensity (A/m)
ρ	volume charge density (C/m^3)
ε_0	permittivity of free space constant equal to $8.85418782 \times 10^{-12}$ m^{-3} kg^{-1} s^4 A^2
μ_0	is permeability of free space constant equal to $1.25663706 \times 10^{-6}$ m kg s^{-2} A^{-2}
σ	is electric conductivity.

With these equations, we can derive all the special case equations we were taught in school for electromagnetism.

Coulomb's Law

As an example, in order to derive the famous Coulomb's Law for a point charge in a vacuum producing an electric field only (i.e., the charge is not moving) we use *Gauss' Law for the Electric Field* equation in the integral form.

This generalized equation is designed to handle any arbitrary surface S from an irregularly shaped object (assuming you can mathematically describe it) and allows us to sum up the electric field **E** on the surface to determine total charge Q with the help of a mathematical solution obtained after solving the integral equation. Or, through the same solution, if we change the total charge of the object Q, we will know the electric field **E**. Either way is fine. But in order to solve the equation, sometimes you need to simplify the situation in order to make the calculations to find a solution simple. Since we are talking about a point charge, let us approximate this situation for the sake of minimizing calculations (and our valuable time) by assuming a perfect sphere of radius *r*.

Thus the equation will be,

$$\varepsilon_o \oint_S \mathbf{E}.d\mathbf{S} = \int_v \rho\, dv$$

For simplicity's sake, we will assume the charge density throughout the volume of the object is the same. Thus ρ is effectively the total charge Q. So we have,

$$\varepsilon_o \oint_S \mathbf{E}.d\mathbf{S} = Q$$

or

$$\oint_S \mathbf{E}.d\mathbf{S} = \frac{Q}{\varepsilon_o}$$

Now the electric field $\mathbf{E}$ for a sphere is constant throughout the surface S since the electric charge is evenly distributed throughout the sphere (including the surface). In fact, using the vector notation $\mathbf{E}$ (in bold) is unnecessary since it is effectively a constant at any point on the surface. Therefore,

$$E \oint_S d\mathbf{S} = \frac{Q}{\varepsilon_o}$$

Finally, integrating dS will give us the total surface area for a sphere. Fortunately, no special tools or university degree is required to calculate this. If we recall from our high school days, the surface area of a sphere is:

$$\oint_S d\mathbf{S} = S = 4\pi r^2$$

Therefore, the solution to the integral equation becomes,

$$E = \frac{Q}{4\pi\varepsilon_o r^2}$$

This is *Coulomb's Law*.

To apply this solution properly to the real world, you can imagine r is the distance from a point charge. In which case, the electric field intensity E will be at a point in space. But since this formula applies equally well to a sphere of radius r with charge Q distributed evenly on its surface, calculating E will come up with the same answer so long as you are aware that the point in space now refers to the surface of the sphere. If you are talking about any position in space above the sphere, you need to subtract the distance *r* from the centre of the sphere by the radius of the sphere. Then you can imagine yourself moving away at a great distance until the sphere looks like a point charge, and the result for E will be the same.

And if another point charge Q_2 suddenly materializes in space, an electrostatic force is exerted on this charge by the other charge through its electric field **E**. The force F can be calculated by,

$$F = Q_2E$$

where E is the electric field intensity calculated for the first charge at a distance r, which just so happens to be the position of the second charge. So now you can calculate the acceleration and velocity at the instant the electric field **E** of the first point charge is exerted on the second charge.

Of course, you can have any kind of irregularly shaped object (and any number of them in space having their own charges) doing all sorts of amazing things and can calculate exactly what this (or any) object will do (i.e., where and how fast at some point in time) using Maxwell's equations and Newton's law of motion. In fact, you can make your life as difficult as you like when calculating the final solution, but you can always choose to simplify the situation at any time. Either find a simpler way to solve your equations, or feel free to use your mind to carefully take into account all the variables and spend time visualizing clearly what is likely to happen in a process known as a *thought experiment*. Often, the latter is the most effective way for dealing with the most complex problems imaginable.

Therefore, whether you need to do mathematical calculations or imagine it is totally up to you.

If it comes to convincing others of your insights, you may need to provide mathematical calculations. Or, you can perform the experiment in real life. Because once people see it and it works as you expected in your own mind, it doesn't matter if people don't have the calculations at hand. Everyone knows it is for real because they can *see* it. Then people can spend all the time in the world developing sophisticated equations to support the observations if they so wish.

Whether you use mathematical equations or perform experiments in your mind and later in real life, either way is perfectly fine in the world of science.

A summary of electromagnetism

Without deriving each and every individual special case equation in electromagnetism, let us present a quick summary of the results.

The fundamental particles of charge

Matter is composed of two fundamental particles[1]: the *electron* and the *proton*. The electric charge of a proton is exactly opposite that of an electron. The charge measured by scientists for an electron is $-1.6021892 \times 10^{-19}$C. For a proton, you simply replace the minus sign with a plus sign to give you the charge.[2]

The electrical unit of charge is one coulomb (C) and is equal to the quantity of charge (since the electron is the one that moves, we tend to use this particle for measuring charge and its flow) passing a point in a conductor in one second when a steady current of one ampere is flowing. We define current as the flow of electric charge. One ampere (1A) is the unit of current equal to the flow of 6.25×10^{18} electrons per second.

A simplified view of an atom consists of a positively-charged nucleus (radius less than about 10^{-14} meters) containing protons (and neutrons when the number of protons is more than one). The nucleus spins, emitting synchrotron radiation, which in turn produces stationary

---

1 The neutron is another stable particle found in nature, but it too is composed of an electron and a proton. Due to the way the electron and proton are combined, the neutron has no apparent electric charge that scientists can detect (although in reality this is probably untrue).

2 Not relevant to this discussion but worth noting is the fact that an electron is about 1/836 times the mass of the proton.

gravity (or radiation) waves in concentric rings around the nucleus of decreasing energy density the further one moves away from the nucleus. These waves keep electrons in position as the particles rapidly orbit the nucleus. The combination of the nucleus and electrons constitutes the atom.

As the atom tumbles around in space, it will look like it is surrounded by a cloud of electrons such that the atom is electrically neutral. But an atom can be electrically charged by adding or removing electrons to create either a negatively or positively-charged atom, respectively. So the total charge Q_1 when N electrons are removed from an atom is $Q_1 = +Ne$, where e is the electric charge for one electron. If other atoms are similarly charged, then the total charge is formed by summing up individual charges Q_1, Q_2, Q_3,...,Q_N to get $Q_{total} = Q_1 + Q_2 + Q_3 + ... + Q_N$.

Under normal circumstances, positive and negative charges are more-or-less spread uniformly within and over a material object, making it, on the large scale, electrically neutral as a whole. However, certain microphysical processes may separate electrical charges to cause one region of the object to have greater negative (or positive) charge than another. When this happens, the object is said to be charged (or electrified).

Electric fields

The study of electrostatics (or static charge physics) is concerned with how the electric field **E**, charge Q, and force **F** are related[3]. Of the three quantities mentioned here, only charge is scalar (that is, it has magnitude but no direction), the others being vectors.

Suppose two objects possess like charges (i.e., two negative or two positive charges) and are brought into close proximity to one another. The result is an electrostatic force between the two that pushes the objects away from each other. For two objects having unlike charges, a force of attraction is experienced instead. This force of attraction or repulsion has a magnitude directly proportional to the product of their charges and inversely proportional to the square of the distance between them.

---

3   Letters that appear in **bold** are called vectors, meaning they contain information about the direction and magnitude).

If we are focussed on just the magnitude (or strength) of the force between the charges, we can derive from Maxwell's equations the following equation[4]:

$$F = \frac{1}{4\pi\varepsilon_o} \cdot \frac{Q_1 . Q_2}{r^2}$$

It is interesting to note that a very similarly structured formula exists for two bodies of mass and the force of gravitation for attracting the bodies as observed by Sir Isaac Newton. It seems the only difference is the magnitude of the forces involved. For example, imagine two bodies of 1 kg each, and both have a charge of +1C and -1C, respectively. We find the ratio of the force of gravitation between the masses and the force of electrostatics between the charges at 1 meter of separation to be about $1:1.347 \times 10^{20}$. The electrostatic force of attraction (and repulsion), therefore, outstrips that of gravitation based on this simple example. Of course, the most interesting question is whether we are dealing with two separate forces of nature. Some traditional scientists think there is a difference purely on the basis that the magnitudes of the forces are so significantly different. But is this really a sufficient argument in the face of what we know from Einstein's Unified Field Theory?

Leaving aside this controversy, the equation we see above is designed to give only an indication of the strength (or magnitude) of the force between charges. But what about direction? To determine the direction as well as the magnitude of the force, we use vectors. Vectors are like numbers, except they carry extra information representing the position in space for an object using x, y and z coordinates, and with it an indication of the direction the object is located compared to another object or point in space. This information is shown as three numbers written like so,

$$\mathbf{A} = (a_1, a_2, a_3)$$

---

[4] The dot (.) symbol between vectors means the vectors must be multiplied using the dot product method. Any dots between scalar quantities or standard international units mean ordinary multiplication.

The first number a_1 shows how far along the x coordinate one should go to represent whatever we are measuring for the object, a_2 is along the y coordinate, and a_3 is along the z coordinate. If you have another vector,

$$\mathbf{B} = (b_1, b_2, b_3)$$

You can apply the simple addition formula to combine these vectors like so,

$$\mathbf{A} + \mathbf{B} = (a_1+b_1, a_2+b_2, a_3+b_3)$$

Assuming the result of this addition gives us the vector,

$$\mathbf{C} = \mathbf{A} + \mathbf{B} = (c_1, c_2, c_3)$$

We can determine the magnitude using the formula,

$$|\,\mathbf{C}\,| = \sqrt{c_1^2 + c_2^2 + c_3^2}$$

So effectively, we get more bang for our buck in terms of extra information about what the objects are doing in space (i.e., its direction) as well as how much force is exerted.

The only problem with vectors is how to write this extra information in an equation without losing the overall structure of the equation. And we all know how much we hate to write highly complex equations containing lots of different symbols, letters and numbers. So, mathematically speaking, to simplify the writing of these three numbers, scientists use a letter to represent a vector in the usual algebraic sense except the letter is given the **bold** style treatment to make this distinction. This means, when calculating the result, a little extra care is needed to manipulate vectors in an equation to get the correct answer.

If a charge Q_1 is placed at a point specified by the vector $\mathbf{r}_1$, and another charge Q_2 at $\mathbf{r}_2$, the force $\mathbf{F}_{1,2}$ acting on the charge Q_1 due to the presence of the second charge can be expressed in vector notation as:

$$F_{1,2} = \frac{Q_1 Q_2}{4\pi\varepsilon_o} \frac{\mathbf{r}_{1,2}}{r_{1,2}^2}$$

where

$$\mathbf{r}_{1,2} = \frac{\mathbf{r}_1 - \mathbf{r}_2}{|\mathbf{r}_1 - \mathbf{r}_2|}$$

is a unit vector and,

$$r_{1,2} = \sqrt{\mathbf{r}_{1,2} \cdot \mathbf{r}_{1,2}}$$

is the magnitude of the distance separating the two charges.

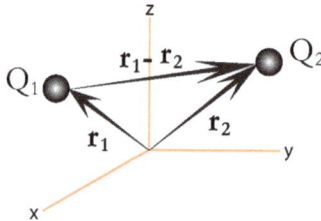

If a group of electrostatic forces are acting on charge Q_1 at $\mathbf{r}_1$ due to other charges Q_2, Q_3, Q_4,...,Q_n, then the resultant force on Q_1 is the sum of all the individual forces exerted on that charge by each of the other charges. Mathematicians write this in equation form as:

$$F_{1,2,...,n} = \frac{Q_1}{4\pi\varepsilon_o}\left(\frac{Q_2 \mathbf{r}_{1,2}}{r_{1,2}^2} + \frac{Q_3 \mathbf{r}_{1,3}}{r_{1,3}^2} + ... + \frac{Q_n \mathbf{r}_{1,n}}{r_{1,n}^2}\right)$$

or the more accepted mathematically shortened form would be,

$$F_{1,2,...,n} = \frac{Q_1}{4\pi\varepsilon_o}\sum_{m=2}^{n} \frac{Q_m \mathbf{r}_{1,m}}{r_{1,m}^2}$$

It should be remembered that it does not matter how big the conductors or insulators carrying the static charge are or what they are made of (or even their masses unless they are really massive), only the magnitude of the charge on them determines the electrostatic force they exert on each other.

302

As a result of this invisible force acting on static charges, a region of space where such (invisible) electrostatic forces are present is called an electrostatic (or static electric) field and is usually illustrated diagrammatically by lines of force with arrows indicating the direction in which a tiny positive charge at that point would move. Negative charges, of course, move in the opposite direction. The closeness of the lines represent the intensity of the field, and if the charged object is made of a purely conducting material, then the lines of force always leave from, or arrive, at right angles to its surface. Electric field lines never cross each other.

Thus, if we want to know the electric field $\mathbf{E}$ at a point in space positioned at $\mathbf{r_0}$ in the presence of a charge Q_1 positioned at $\mathbf{r_1}$, we would write the formula as:

$$\mathbf{E}_{0,1} = \frac{Q_1}{4\pi\varepsilon_o} \frac{\mathbf{r}_{0,1}}{r_{0,1}^2}$$

So when you multiply Q_1 with $\mathbf{E}$, you get the force, $\mathbf{F}=Q_1\mathbf{E}$ (a bit like Newton's formula $\mathbf{F}=m\mathbf{a}$).

Similarly, the electric field at some specified point $\mathbf{r_0}$ in space due to a number of point charges Q_1, Q_2, Q_3,...,Q_n is simply the vector addition of the fields due to each individual charge. In other words,

$$\mathbf{E}_{0,1,...,n} = \frac{1}{4\pi\varepsilon_o} \sum_{m=1}^{n} \frac{Q_m \mathbf{r}_{0,m}}{r_{0,m}^2}$$

Another thing often talked about in electromagnetism is the concept of potential difference, or voltage.

Voltage is a scalar quantity showing the difference in electrical potential energy between two points per unit charge. Expressed in another way, it is the work done in moving one coulomb of charge from one point to another. The standard unit for voltage is the volt (1V), equivalent to one joule per coulomb ($1J.C^{-1}$).

The potential difference at a distance r from a point charge Q_1 is given by:

$$V = \frac{Q_1}{4\pi\varepsilon_o r}$$

As for an arbitrary collection of charges in space, the formula becomes the sum of the voltages produced by each individual charge like so:

$$V = \frac{1}{4\pi\varepsilon_o} \sum_{m=1}^{n} \frac{Q_m}{r_m}$$

If the electric field **E** is specified but not the charges, the voltage can still be worked out directly from the field using the line integral,

$$V = \oint_L \mathbf{E}.d\mathbf{L}$$

Magnetic fields

Magnetic phenomena occur naturally in iron ores. The Chinese were probably the first to make use of the magnetic properties of iron as early as CE 121 to determine roughly where north and south are (since the magnetic poles of the Earth are very close to the geographic poles of the planet). This was done by suspending an iron rod, which, when allowed to move on its own, will line up with the Earth's natural magnetic field lines.

When drawing the lines of a magnetic field on paper, an arrow on the line would represent the direction the North pole of a compass needle would point. And like the electric field, magnetic field lines never cross each other.

Now imagine a charge Q moves with velocity **v** in the presence of an external magnetic field only. It begins to sense another force coming into play. As the charge moves through the magnetic field of strength **B**, a magnetic force **F** is applied on the object at right angles to the direction of motion and to the magnetic field. The magnitude of the force is given by,

$$F = Q.v.B.\mathrm{Sin}\beta$$

where β is the angle at which the charged object first enters the field. In vector notation, the direction and magnitude of this magnetic force is given by,

$$F = Q(v \times \mathbf{B})$$

where $v \times \mathbf{B}$ is the cross product of two vectors.

If the direction of vector v for the charge happens to be slicing through the outer magnetic field perpendicularly, then the magnitude of the force is maximal and equal to,

$$F = Q.v.B$$

The only situation where the charge does not get deflected by the magnetic force is if it enters the field in the same direction as the field itself.

As a consequence of the way the magnetic field exerts a force on the charge at right angles to its direction of motion, it cannot alter the magnitude of the charged object's velocity using this field alone. Any variation in velocity requires the existence of another external magnetic field to interact with the charge and whose source of field generation must be physically joined to this particle. However, under the action of a single magnetic field only, a charged object in motion through this field is coerced to move in circular paths of radius R. The circulating charge carries current around in a circle in such a direction that the magnetic field strength within the circle is diminished despite the potentially high magnetic intensity that may exist outside the circle.

To calculate R, we use the centripetal force equation:

$$F = \frac{m.v^2}{R}$$

and combine it with the maximal magnetic force equation like so:

$$\frac{m.v^2}{R} = Q.v.B$$

Making R the subject of the equation, we get,

$$R = \frac{m.v}{Q.B}$$

If charges are confined to a long thin conductor that is moving through the external magnetic field, the **v** x **B** term can be considered equivalent to an electric field **E**. However, this **E** field induced by the magnetic field **B** inside the conductor is of non-electrostatic origin since it can penetrate matter even when there is no current. What this field actually is has been a matter of some debate and has been of interest to a number of scientists, including electrical engineer Dr. Wilbert B. Smith in the late 1950s, but for the purposes of this section, we shall call this field the *motional electric field*. Its presence inside the conductor will simply separate the charges to opposite ends of the conductor.

The standard unit of measuring the intensity of a magnetic field is called one Tesla (1T), which is equivalent to 1 newton per ampere per meter ($1N.A^{-1}.m^{-1}$). Sometimes the Gauss (1G) is used to measure the magnetic field but this is only useful for weak magnetic fields. Thus, $1T=10^4G$.

Laboratory-made static magnetic fields can reach as high as 30T, while some pulsed-current electromagnets can generate magnetic fields up to 120T in a time interval of about one millisecond. In contrast, the Earth's magnetic field intensity at its surface is very weak and is of the order of 0.000075T (or 0.75G) at the mid-northern and mid-southern latitudes, and about 0.31G at the equator. Despite its low value, the magnetic field of the Earth is sufficient to deflect many high-speed charged particles emitted by the Sun that can harm much of life on this planet.

Speaking of the Earth's magnetic field, the converging nature of the field reaching two points on the surface called the *magnetic poles* helps to create three types of periodic motion for an electric charge entering this field:

1. Spiraling motion.
2. Bounce motion.
3. Drift motion.

In the first case, charged particles gyrate (or spiral) around magnetic field lines. The gyrofrequency f of the motion is given by,

$$f = \frac{B.Q}{2\pi m}$$

where B is the magnetic field strength, Q is the charge of the particle in motion, and m is the mass of the particle. For example, an electron with a charge of $-1.6021892 \times 10^{-19}$C and a mass of 9.10939×10^{-31}kg passing through a magnetic field of strength 10^{-9}T will spiral around the magnetic field lines (in an anti-clockwise direction) with a gyrofrequency of about 2,800 times per second. A proton, on the other hand, may have the same electric charge but with a positive sign, so the spiraling motion will go the opposite way, but its mass of 1.6726×10^{-27}kg is much heavier. This means the proton will move more slowly through the same magnetic field than the electron. Thus, the gyrofrequency for the proton will be about 1.5 times per second.

Furthermore, as the charge spirals around the magnetic field lines, electromagnetic radiation is emitted by the charge outwards in a narrow cone because of its continual changes in motion. The wavelength of this emitted (synchrotron) radiation gets shorter as the electron travels faster.

In the second case, charged particles perform what is known as a "bounce" motion at a point where the magnetic field lines converge. The point of convergence is called a mirror point. When a charged particle moves from a region of low magnetic field strength to a region of higher magnetic field strength, two things happen to the motion of the particle:

1. The gyroradius decreases.
2. The forward motion is diminished.

So, when a particle reaches a mirror point, the particle has no forward velocity, and since the same forces that decelerated the forward motion are still acting on the particle, it reflects away from this point and starts to accelerate until the magnetic field lines converge once again.

In the third case, a charged particle bouncing back and forth along a *curved* magnetic field line will experience drift motion. Despite the fact that protons and electrons gyrate in opposite directions around a magnetic field line, all electrons and protons drift in the same direction and at the same velocity.

Looking at moving charges inside a conductor, it is known that when current I flows at a constant rate through a conducting wire (represented by vector **L** running along the wire and pointing in the direction of the current), a magnetic field is generated. When this wire, with its own magnetic field, is placed inside another magnetic field **B**, the two fields interact, and a magnetic force **F** results, which pushes the wire away. The formula that brings all these quantities together is,

$$\mathbf{F} = I\,(\mathbf{L} \times \mathbf{B})$$

Should the reader be more concerned with just the magnitude of the force, it may most easily be obtained from the formula,

$$F = I.L.B.\mathrm{Sin}\beta$$

where L is the length of the wire immersed in the external magnetic field, and β is the angle the conductor makes with the field.

In another special case situation, two long, straight conductors of length L, separated by a distance d, and carrying current I_1 and I_2 in the same (or opposite) direction, will experience a force of attraction (or repulsion) because each conductor lies in the magnetic field set up by the other. The magnitude of the magnetic force so generated between them is given by,

$$F = \mu_o\,\frac{I_1.I_2.L}{2\pi.d}$$

In this equation, μ_o is the permeability of free space (also known as the magnetic constant).

It should be noted that in some materials, the permeability can be higher than in others. If this is the case, this simply means that some materials are able to concentrate and strengthen a magnetic field more than others. So if the wires are immersed in a different material that is not a vacuum, then μ_o is substituted for μ and calculated using the equation,

$$\mu = \mu_o . K_m$$

where K_m are numerical values obtained by the physicists for different materials (you need to check a reference book).

Type of material	Value of μ	Value of K_m
Vacuum	μ_o	1
Paramagnetic	Slightly larger than μ_o	Slightly larger than 1
Ferromagnetic	Much larger than μ_o	Much larger than 1
Diamagnetic	Slightly smaller than μ_o	Slightly smaller than 1

The same sort of thing also exists for the dielectric permeability of a material, which means that some materials have different abilities of concentrating the electric field.

To calculate the magnetic force exerted on a charge Q_2 moving at velocity v_2 due to the magnetic field generated by the velocity v_1 of another charge Q_1, the equation is,

$$F = \frac{\mu_o}{4\pi}\left(\frac{Q_1 . Q_2}{r^2}\right) v_2 \times (v_1 \times r)$$

where the scalar quantity r is the distance separating the two charges, which can be calculated from the vector **r** via,

$$r = |r| = \sqrt{r.r}$$

Finally, but by no means the end (indeed, you can spend lots of time generating heaps of magnetic formulas for every conceivable special case situation you can imagine until the cows come home), you can calculate the magnetic field **B** at any point space where there is a charge Q moving at constant speed. The charge has a velocity of **v** (in the vector notation for determining the direction of the charge). The equation is,

$$\mathbf{B} = \frac{\mu_o}{4\pi}\left(\frac{Q}{r^2}\right)(\mathbf{v} \times \mathbf{r})$$

Electromagnetic fields

The final aspect of electromagnetic theory is what happens when a magnetic field changes its strength in a non-static way, or if a conductor accelerates or decelerates through the magnetic field. In this situation, the field somehow induces a motional electric field in space or within the conductor. This induced field, or the combination of the two fields, somehow affects charged (and apparently uncharged) objects. This is called *magnetic induction*. Scientists then call the resultant field an *electromagnetic field*.

When the motional electric field is created in space by a time-varying magnetic field, the magnitude of this motional electric field is given by,

$$\sqrt{\varepsilon.\mu}$$

where ε and μ are the dielectric permittivity and magnetic permeability of a material, respectively. In a vacuum, the speed of light can be calculated as,

$$c = \frac{1}{\sqrt{\varepsilon_o.\mu_o}}$$

And if you ever need to calculate the voltage and current in a conductor accelerating (or decelerating) in a magnetic field, or a stationary conductor immersed in an oscillating magnetic field (i.e., an electromagnetic field), you will find numerous ready-made equations in any good physics textbook.

For a quick summary of electromagnetic theory, this is probably all you need to know.

More about vectors

Let us restrict our attention to vectors in three dimensions.

A vector is defined as a quantity with magnitude and direction.

For those readers not familiar with dot product and cross product of two vectors, suppose the vectors **A** and **B** are defined as:

$$\mathbf{A} = (a_1, a_2, a_3)$$
$$\mathbf{B} = (b_1, b_2, b_3)$$

The cross product of **A** and **B**, defined as **A** x **B**, is,

$$\mathbf{A} \times \mathbf{B} = (a_2.b_3 - a_3.b_2,\ a_3.b_1 - a_1.b_3,\ a_1.b_2 - a_2.b_1)$$

The dot product of **A** and **B**, defined as **A.B**, is,

$$\mathbf{A.B} = a_1.b_1 + a_2.b_2 + a_3.b_3$$

Note that the dot (.) means you multiply the numbers together. Subtraction and addition of vectors is straightforward:

$$\mathbf{A} - \mathbf{B} = (a_1 - b_1,\ a_2 - b_2,\ a_3 - b_3)$$

$$\mathbf{A} + \mathbf{B} = (a_1 + b_1,\ a_2 + b_2,\ a_3 + b_3)$$

$$|\mathbf{A}| = \sqrt{a_1^2 + a_2^2 + a_3^2}$$

And the magnitude of a vector is calculated by the following equation:

APPENDIX C

The Origin of Science

I n the sixth century BCE, a group of curious Greek people called the "hylozoists" supported a single great religion. Evidence of this can be found in some of the writings of the Greek philosopher Heraclitus of Ephesus (540–c.480 BCE). What little we have left of this philosopher's knowledge, covering a little more than a hundred sentences, nonetheless shows considerable understanding of certain concepts espoused by this religion, including the idea of "everything is in flux" (or in a state of constant change) and the notion of the "unification of opposites".

In the former statement, Heraclitus effectively assets that all matter is alive and in possession of inherent latent, but perhaps not conscious, powers.

In the latter statement, Heraclitus observes the presence of opposites and argues that the continuous cyclic movement and balance between them are essential for the existence of life and the Universe. Rather than surviving independently, these opposites rely on each other for their shared existence through a system of balanced exchanges.

Once Heraclitus understood the unified nature of these opposites when achieving balance and realized that they were essentially present in all things, he gave his undivided support for the unity of the Universe, stating, "All things are one". Thus, despite the passing of more than 2,500 years, these ancient philosophical views from Greece promoted by the hylozoists and Heraclitus still ring true among Eastern mystics.

Greek philosopher, Heraclitus.

However, in the fifth century BCE, this single great religion was disrupted by an influential philosopher named Parmenides of Elea and others whose views[1] conflicted with those of Heraclitus. Parmenides and others supported a non-unified view of opposites, in the sense that the opposites had no obvious connection and, therefore, existed for other reasons. As Parmenides states:

---

[1]   For instance, the fundamental understanding that Parmenides and others had of the nature of matter was that it was indestructible and made of the tiniest "basic building blocks" (later called "atoms") that were intrinsically devoid of life and moved according to some external force they had to assume was of spiritual origin.

"Either a thing is, or it is not."

In other words, an opposite must exist or else it does not, and the state of one opposite at a particular moment in time has no influence on that of the other. For instance, Parmenides did not believe that creation (one opposite) and destruction (another opposite) could exist simultaneously; hence, no unification of the opposites was possible.

This "fragmented" view led to a separation of matter and spirit, animate and inanimate, life and death, and so on. As a result of this "as we directly see it in an instant in time" view, two new religions took its place. One of the religions led to Eastern mysticism, and the other to a Western religion that would eventually introduce science to the modern world.

Following this great flurry of intellectual and cultural activities in ancient Greece, Western religion turned its attention to describing and gathering knowledge about the spiritual world. It was then that Aristotle systematically compiled and organized this part of the knowledge. He did not include knowledge of the material world, deeming it unimportant in the face of ideas relating to God, the soul, and human ethics. Once the ideas were brought together, they would be cemented beyond further scrutiny or dispute through a powerful European monarchy—the Church. For centuries, these ideas would dominate Western thought.

In the 16th century, Western religion experienced challenges and eventually faced significant advancement when scholars learned to be inquisitive about the material world. One such person was Galileo Galilei, an Italian who became the first person to combine the emerging subject of mathematics with observations regarding the material world. However, the view of nature held by Galileo (as well as various other rebellious individuals commonly known as scientists) still had a distinctly "fragmented" flavor about it.

Another important historical figure who fervently upheld this "fragmented" belief was René Descartes. Descartes, often regarded as the founder of modern philosophy, asserts that:

> "There is nothing included in the concept of body that belongs to the mind, and nothing in the mind that belongs to the body."

He thus calls the mind the "thinking thing" and matter the "extended thing" as part of his method of fragmenting the mind and body.

The "fragmented" approach strongly influenced Western scientists such as the highly distinguished astronomer and mathematician Sir Isaac Newton. The man who was fabled to have been hit on the head by an apple from a tree and wondered how things could fall to the ground with such ease, Newton viewed the material world as a great machine with many lifeless parts, each of which can be controlled and exploited individually at will by anyone without being affected by the other parts. The result, which endured until the beginning of the 20th century, was a purely materialistic and mechanistic Newtonian model of the Universe, which was totally devoid of "spiritual" and emotional values, especially when referring to people. Hence, all things such as feelings and holistic thinking (or intuition) were cast out of the realm of modern scientific discourse.

Not so among Eastern mystics. Here, the mystics believe that this "fragmented" approach[2] to reality is the cause of the many social, economic, and environmental crises we see today. This fragmentation approach may provide certain advantages for some people; for instance,: we now have technology and classical physics to explain the material world. However, this approach has been detrimental in that nations are divided; natural resources are not distributed equitably; political and economic systems are dividing people and their views; the environment is being frivolously polluted and destroyed; people's emotions are not considered in any fragmented solution; and the social deterioration caused by fragmentation has incited waves of violence, mental illness, and apathy. This is all attributable to our belief that the perceived world should be divided into separate and unrelated things and events for us to categorise, measure, and treat according to our own values and beliefs.

Eastern mystics indicate another way to approach reality, namely the holistic approach. According to this view, rather than divide the world into individual, lifeless and unrelated parts, we are told that true stability, long-term happiness, and greater meaning to life comes when we see the interconnectedness of the world. To achieve this, it is

---

2   And controlled in a hierarchical manner by males over a long period of time.

essential for people to uncover for themselves the global "unifying" patterns connecting all things and events, and consequently realizing how these factors can affect one another in essential ways. Only when the link is properly understood can we find solutions and achieve greater things together.

These mystics have compiled in support of the concept of the unity (or holistic approach) of opposites (the fragmented approach) in the Universe the following beliefs:

1. All things and events in the Universe are "alive";
2. All things and events in the Universe are in a state of continuous motion; and
3. All things and events in the Universe are interrelated and inseparable, both spiritually and materially, at the same time.

It is claimed that this awareness of reality—known to mystics as "enlightenment"—is a physical experience as well as an intellectual act, in which the person's entire being (both mind and body) is unified and involved.

Surprisingly, this holistic knowledge has been acquired in essentially the same way that Einstein acquired his knowledge when studying the problem of high-speed physics. Mystics employ the technique of combining their visualization and creative skills with direct observations of the universe (using all their senses) to help them identify in the mind certain hidden large-scale and potentially "unifying" patterns. Similarly, Einstein applied the same visualization and creative skills to his work since he understood that no machine existed that could help him to observe what happens when one approaches the speed of light. Instead, he used his mind to imagine, predict the possible outcomes, and apply his well-developed rational mind based on all the information he gathered from the available scientific sources. Then, with enough visualization and careful application of mathematics, Einstein was able to develop a revolutionary new idea that would transform the world of physics.

While Einstein used his visualization and creative skills to focus on one aspect of physics, the same skills can be applied on a larger scale to identify grander patterns in everything that surrounds us. Some of the results of performing this work can be seen in the above observations compiled by Eastern mystics in support of the concept of unity (and dualism, if we want to include the paradoxical property of God).

Then, in the 1920s, quantum theory took the scientific world by storm. Scientists soon noticed a link between the ideas contained in quantum theory and Eastern mysticism; for instance, the concept of continuous motion and change, which is described in Eastern mysticism in addition to hylozoist writings, is well-supported by quantum scientists. Fritjof Capra, author of *The Turning Point* and the *Tao of Physics*, states that:

> "According to quantum theory, matter is always restless, never quiescent."

Einstein also supports this view, asserting that there is no absolute frame of reference in the real Universe that has zero velocity (or even absolutely uniform motion) with respect to everything that the Universe contains.[3]

Similarly, Dr. Henry Stapp of the University of California confirms the idea that on the atomic level, all particles and events are interconnected. Stapp states that:

> "An elementary particle is not an independently existing unanalyzable entity. It is, in essence, a set of relationships that reach outward to other things."[4]

As for whether everything is "alive", a piece of rock or metal may seem "dead" on the macroscopic level, but by magnifying such a material, one can see the hustle and bustle of activity associated with atoms and molecules. The more closely we look, the more alive the seemingly inanimate object appears to be.

That could explain why Russian biochemist Dr. Alexander Ivanovich Oparin (1894–1980), a former professor at Moscow University, said as early as 1924:

> "There is no fundamental difference between a living organism and lifeless matter. The complex combination of

---

3   Bernstein 1991, p.41.
4   Capra 1991, p.151.

manifestations and properties so characteristic of life must have arisen in the process of the evolution of matter."[5]

There are definite parallels between Eastern religious views and Western scientific views, as American physicist Dr. Julius Robert Oppenheimer (1904–1967) notes:

"The general notion about human understanding... illustrated by discoveries in atomic physics [is] not in the nature of things wholly unfamiliar, wholly unheard of, or new. Even in our culture [such ideas] have a history, and in Buddhist and Hindi thought a more considerable and central place. What we shall find is an exemplification of old wisdom."[6]

Danish physicist Niels Bohr, who contributed significantly to the study of atomic physics and quantum theory, similarly indicates a connection between science and religion:

"For a parallel to the lesson of atomic theory [we can turn to] ... problems with which already thinkers like Buddha and Lao Tzu have been confronted."[7]

And now, for the first time, science can apparently use Einstein's Unified Field Theory to highlight yet another interesting piece in the great puzzle when revealing the connection between religion and science. Specifically, we have identified a paradoxical entity in science that matches remarkably well with another paradoxical entity, namely God, in religion.

Controversial? Hell, yes! Should we apologize for it? Of course not. Will scientists be happy? Probably not. There will always be numerous scientists who will prevaricate the "God issue" and endeavor to prove that their work has absolutely no relationship to this mysterious entity. For example, Peter Atkins, an English chemist and professor of chemistry at Lincoln College, Oxford, England, was asked by Rod Liddle in the documentary *The Trouble with Atheism*, "Give me your views on the existence, or otherwise, of God". Atkins responded by saying,

---

5   Edelson 1980, p.33.
6   Oppenheimer 1954, pp.8-9; Capra 1991, p.321;
7   In a speech on quantum theory delivered in October 1937 at *Celebrazione del Secondo Centenario della Nascita di Luigi Galvani*, Bologna, Italy.

"Well it's fairly straightforward: there isn't one. And there's no evidence for one, no reason to believe that there is one, and so I don't believe that there is one. And I think that it is rather foolish that people do think that there is one."[8]

But as Einstein once said,

"Then there are the fanatical atheists whose intolerance is the same as that of the religious fanatics, and it springs from the same source . . . They are creatures who can't hear the music of the spheres."[9]

---

8   *The Trouble with Atheism*. UK Channel 4 TV, televised on December 18, 2006.
9   Einstein replying to a question from an unidentified person on August 7, 1941. Recorded in the
         Einstein Archive, reel 54-927.

Glossary

A.D.: Acronym for "After Death" used by traditional historians as a reference point for when events in history have taken place after the death of Jesus. However, when combined with the acronym B.C. (meaning "Before Christ"), which is the reference point for when Jesus was born, people can get confused as to how to specify a year for an event that took place during the life of Jesus. Furthermore, the acronyms are seen as supporting Christianity and not other religions. To avoid this situation, it is more common nowadays to use CE for "Current Era" or BCE. for "Before Current Era" such that the birth of Jesus is set at 4 BCE and his death at CE 33, and events in between these years can be identified. This makes the reference point for our modern calendar system start at 0 with no connection to any religious event or person.

Atom: A microscopic object consisting of electrons, protons, and neutrons. The atom is structured such that the electrons move at high speeds in discrete orbits around a fast-spinning nucleus containing protons and neutrons.

Atomic: On the scale of the atom.

B.C.: Acronym for "Before Christ". This means the reference point starts at when Jesus Christ was born, but this has now been replaced by BCE.

B.C.E: Acronym for "Before Current Era". This means the reference point starts at year 0 and covers all events that took place before this starting point.

C.E.: Acronym for "Current Era". This means the reference point starts at year 0 and covers all events that took place after this starting point.

Conductor: A material that allows the rapid movement of electrical, heat, or sound energy through the material.

Current, electric: The flow of electric charge.

Doppler effect: An apparent variation in the frequency of a wave (light or sound), owing to the motion of the wave source relative to the listener/observer.

Electric field: A region of space where invisible electric forces are present. Affects only charged objects.

Electromagnetic field: Electric and magnetic fields. Static electromagnetic fields are believed to affect only charged objects. Oscillating electromagnetic fields (also called radiation) affect both charged and uncharged objects.

Electromagnetic waves: Regions of intense electromagnetic energy called photons, which move at the speed of light. Also called radiation.

Electromagnetism: The study of electric charge and the forces that exist between moving and non-moving bodies of varying or non-varying charge. The particle mediator for all the forces of electromagnetism is the photon.

Energy: That which is carried by electromagnetic waves by way of electromagnetic fields in order to create matter and all the forces of nature.

Frequency: The number of oscillations (or cycles) in a wave in one second, measured in hertz (Hz).

General Theory of Relativity: A theory devised by the late Dr. Albert Einstein. The theory is an extension of the Special Theory of Relativity to take into account the accelerating motion of matter and, at the same time, to show how acceleration and gravity are linked.

Gravitational field: A region of space where invisible gravitational forces are present. According to the Unified Field Theory, this is believed to be nothing more than electromagnetic radiation (or photons) moving uncharged matter, including itself.

Inertia: The pressure exerted by electromagnetic radiation on an accelerating body to oppose its tendency to move.

Insulator: A material that inhibits the passage of electrical, heat, or sound energy through it.

Light bending: Predicted by Albert Einstein's General Theory of Relativity and experimentally verified by astronomers, notably Professor Arthur Stanley Eddington during a solar eclipse in 1919. It is caused by light passing through a strong gravitational field.

Light-year: The distance traveled by a ray of light in one year, or 9.454255×10^{15} meters.

Magnetic field: A region of space where invisible magnetic forces are present. Affects only charged objects.

Mass: A region of concentrated electromagnetic and gravitational energy (or essentially just electromagnetic energy according to the Unified Field Theory).

Momentum: A measure of the force or energy of a particle conserved in all collisions with other particles in a closed system.

Newton's Laws of Motion: Famous English mathematician, physicist and astronomer Sir Isaac Newton (1642–1727) established three laws of classical science to describe the observed behavior of bodies with mass. The first law states that all bodies have the observed tendency to remain at rest or to continue in motion in a straight line at constant speed until acted upon by a force. The second law states that a body subjected to the actions of a force will accelerate (or decelerate), according to the differential equation:

$$a = \frac{dv}{dt} = \frac{F}{m}$$

where a is acceleration, F is force, m is mass, and dv/dt is the instantaneous velocity v at any specified time t. The acceleration of the body is in a direction governed by the force. In the third law, any body under the action of a force will exert an equal but opposite force, owing to the inertial tendency of the body to remain at rest or move at constant speed.

Nuclear fusion: The fusing together of two or more atomic nuclei to form a new nucleus and, in the process, emit vast amounts of energy into the surroundings.

Oscillation: A physical quantity of which the value varies over time in a periodic way.

Photon: A region of concentrated electromagnetic energy. A stream of photons propagating through space is described mathematically by scientists as an oscillating electric and magnetic field at right angles to each other and to the direction of motion, also called an electromagnetic wave. In classical physics, a photon is just another form of ordinary matter, albeit invisible.

Radioactivity: The spontaneous emission of radiation from the nucleus of an atom. The radiation emitted is electromagnetic by way of photons, but scientists will broadly expand this to include the helium nuclei, protons, neutrons, and electrons.

Relativistic: Of, or pertaining to, high speeds approaching that of light and that follow the laws of the Special and General Theory of Relativity and of the Unified Field Theory as proposed by the late Dr. Albert Einstein.

Science: A body of useful and practical knowledge and a method of obtaining it. The knowledge epitomizes the experiences of many people and consists of a set of beliefs or theories that help represent and explain the fundamental and reproducible laws of nature under standard conditions.

Special Theory of Relativity: A theory devised by the late Dr. Albert Einstein. The theory discusses what happens to the time, mass and length of an object as perceived (and measured) by an outside observer when the object travels at a very high speed approaching that of light relative to the observer. The result is that time dilates, mass increases, and length contracts for the object as it approaches the speed of light.

Speed of light: Exactly 299,792,458 meters per second.

Stars: Large luminous bodies of hot compressed gases composed mostly of hydrogen and helium, with some heavier elements. Stars form in space like raindrops do on Earth.

Subatomic: On the scale of fundamental particles, such as the electron and proton.

Temperature: A measure of how hot or cold something is. This sensation of hotness or coldness is the result of the quantity and frequency of electromagnetic radiation present in a substance.

Unified Field Theory: A theory devised by the late Dr. Albert Einstein. The theory is an extension of the General Theory of Relativity to take into account the motion of the electric charge and the presence of the electromagnetic field. Indeed, it is believed that the electromagnetic energy density is what actually controls the gravitational field, and density is controlled by the electric charge and the frequency of the oscillation of the electromagnetic field. Or, in classical terms, photons are in many ways just like ordinary matter, and this includes having their own gravitational field. This raises the question as to whether we should maintain the concept of a gravitational field or get rid of the field altogether and imagine photons (or electromagnetic radiation) doing all the work of the gravitational field.

Wavelength: The distance of one repeating unit in a wave. The repeating unit may begin and end at the crests or troughs or any adjacent point of equal phase in a wave.

Weight: A measure of the force exerted by a gravitational field on an object. According to the Unified Field Theory, this is really how much of a radiation shielding effect an object is able to create to help control the radiation pressure exerted by the universal background radiation on the object.

Bibliography

Batten, Mary. "Aging and You - Understanding Life Spans": *Omega Science Digest*. May/June 1984, pp.38-43 & 57.

Benzer, Seymour; Lin, Yi-Jyun & Seroude, Laurent. "Extended Life-Span and Stress Resistance in the Drosophila Mutant": *Science*. October 1998. Volume 282 Issue 5390, pp.943-946.

Berlitz, Charles & Moore, William. 1979, *The Philadelphia Experiment*. London: Granada Publishing Limited.

Berlitz, Charles & Moore, William. December 1991, *The Philadelphia Experiment*. Eighteenth Printing. New York: Ballantine Books.

Bernstein, Jeremy & Kermode, Frank (editor). 1991, *Einstein* (Second Edition). London: Fontana Press.

Bernstein, Jeremy. 1997, *Albert Einstein and the Frontiers of Physics*. New York, USA: Oxford University Press.

Boslough, John. "Searching for the Secrets of Gravity": *National Geographic*. May 1989, pp.563-581.

Burnie, David. 1992, *Collins Eyewitness Science: Light*. Pymble, NSW: Harper Collins Publishers (Australia) Pty Ltd.

Calder, Nigel. 1983, *Einstein's Universe: The Layman's Guide*. New York: Penguin Books.

Capra, Fritjof. 1984 & 1991, *The Tao of Physics*. First & Third Editions. London: Harper Collins Publishers.

Challoner, Jack. 1993, *Collins Eyewitness Science: Energy*. Pymble, NSW: Harper Collins Publishers (Australia) Pty Ltd.

Childress, D. Hatcher (compiler). 1990, *The Anti-Gravity Handbook.* Stelle, Illinois, USA: Adventures Unlimited Press.

Childress, D. Hatcher (compiler). 1990, *Anti-Gravity and the Unified Field.* Stelle, Illinois, USA: Adventures Unlimited Press.

Cooper, Christopher. 1992, *Collins Eyewitness Science: Matter.* Pymble, NSW: Harper Collins Publishers (Australia) Pty Limited.

Couper, Heather & Henbest, Nigel. 1997, *Big Bang: The Story of the Universe.* Sydney, Australia: Harper Collins Publishers.

Cwiklik, Robert. 1987, *Albert Einstein and the Theory of Relativity.* New York: Barron's Educational Series, Inc.

Davies, P.C.W. 1987, *Quantum Mechanics.* London: Routledge & Kegan Paul Limited.

Davies, Paul. 1990, *God and the New Physics.* London: Penguin Books Limited.

Davies, Paul. 1992, *The Mind of God: Science and the Search for Ultimate Meaning.* London: Penguin Books Limited.

Deane, Professor Roger et al. "A Close-Pair Binary in a Distant Triple Supermassive Black Hole System": *Nature.* July 3, 2014. Volume 511, pp.57-60.

Dellinger, Ryan W.; Santos, Santiago R.; Morris, Mark; Evans, Mal; Alminana, Dan; Guarente, Leonard; and Marcotulli, Eric. "Repeat dose NRPT (nicotinamide riboside and pterostilbene) increases NAD^+ levels in humans safely and sustainably: a randomized, double-blind, placebo-controlled study": *NPJ Aging and Mechanisms of Disease.* November 24, 2017.

Devonshire, Hilary. 1991, *Science Through Art: Light.* London: Franklin Watts.

Deyo, Stan. 1992, *The Cosmic Conspiracy.* Kalamunda, Western Australia: West Australian Texas Trading.

Dyson, Sir Frank W., Eddington, Professor Arthur S. & Davidson, Charles R. "A Determination of the Deflection of Light by the Sun's Gravitational Field, from Observations Made at the Total Eclipse of May 29, 1919": *Philosophical Transactions of the Royal Society A: Mathematical, Physical and Engineering Sciences.* Volume 220

(January 1, 1920), pp.291-333. ISSN 1364-503X.
https://www.worldcat.org/issn/1364-503X.

Einstein, Albert. "Zur Elektrodynamik bewegter Korper" (translated as "On the Electrodynamics of Moving Bodies"): *Annalen der Physik*. 4th Series, 1905, Volume 17, pp.891-921.

Einstein, Albert. "On the Relativity Principle and the conclusions drawn from it": Johannes Stark's *Yearbook of Radioactivity and Electronics*. 1907.

Einstein, Albert. "Die Grundlage der allgemeinen Relativitäts-theorie" (translated as "The Foundation of the General Theory of Relativity"): *Annalen der Physik*. 4th Series, 1916, Volume 49, pp.769-822.

Einstein, Albert. "Neue Moglichkeit Fur Eine Einheitliche Feldtheorie von Gravitation Und Electrizitat" (translated as "New Possibility for a Unified Field Theory of Gravitation and Electricity"): *Sonderabdruck aus den Sitzungsberichte der koniglich preussische Akademie der Wissenschaften. XVIII. Phys-Math Klasse.* Berlin: Verlag der Akademie der Wissenschaften in Commission Bei Walter de Gruyter u Co., 1928.

Einstein, Albert. "Zur Einheitliche Feldtheorie" (translated as "About the Unified Field Theory"): *Sonderabdruck aus den Sitzungsberichte der koniglich preussische Akademie der Wissenschaften. I. Phys-Math Klasse.* Berlin: Verlag der Akademie der Wissenschaften in Commission Bei Walter de Gruyter u Co. 1929, pp.2-7.

Einstein, Albert. "Einheitliche Feldtheorie Und Hamiltonsches Prinzip" (translated as "Unified Field Theory and Hamiltonian Principles"): *Sonderausgabe aus den Sitzungsberichte der preussische Akademie der Wissenschaften. X.* Berlin: Verlag der Akademie der Wissenschaften in Commission Bei Walter de Gruyter u Co., 1929, pp.2-7.

Einstein, Albert. "Zwei Strenge Statische Lösungen der Feldgleichungen der Einheitlichen Feldtheorie" (translated as "Two Strict Static Solutions to the Field Equations of the Unified Field Theory"): *Sonderausgabe aus den Sitzungsberichte der preussische Akademie der Wissenschaften. VI.* Berlin: Verlag der Akademie der Wissenschaften in Commission Bei Walter de Gruyter u Co., 1930, pp.2-13.

Einstein, Albert. "Sytematische Untersuchung über Kompatible Feldgliechungen, welche in Einem Riemannschen Raume mit Fernparallelismus Gesetzt Werden Konen" (translated as "Systematic Study of Compatible Fields, which are set in Riemannian Space with Distant Parallelism"): *Sonderausgabe aus den Sitzungsberichte der preussische Akademie der Wissenschaften. XIII.* Berlin: Verlag der Akademie der Wissenschaften in Commission Bei Walter de Gruyter u Co., 1931.

Einstein, Albert. 1934, *Essays in Science.* New York: Philosophical Library.

Einstein, Albert. "How I Created the Theory of Relativity": *Physics Today.* 1982, Volume 35, Number 8, pp.45-47.

Einstein, Albert. 2005, *Mein Weltbild* (translated as "My World View"). Reprint of the unabridged and translated into English. New York: Europa Verlag, Zurich.

Eyewitness Encyclopedia of Science 2.0 interactive CD. 1997, London: Dorling Kindersley Multimedia.

Fairbridge, Rhodes W. (General Consultant) 1988, *Reader's Digest Marvels and Mysteries of the World Around Us.* London: The Reader's Digest Association Limited.

Farmello, Graham. 1982, Unit 13: *The Beginnings of Modern Atomic Physics.* England: The Open University Press.

Fitzmyer, Joseph A. "Did Jesus Speak Greek?": *Biblical Archaeology Review.* 18/5. September/October 1992, pp.58-63.

Fradin, Dennis. 1987, *A New True Book: Radiation.* Chicago, USA: Regensteiner Publishing Enterprises, Inc.

Frank, Philipp. 1948, *Einstein: His Life and Times.* London: Jonathan Cape.

Gamow, George. "Gravity": *Scientific American.* March 1961, Volume 204, Number 3, pp.94-106.

Gamow, George. "The Principle of Uncertainty": *Scientific American,* January 1958, Volume 198, Number 1, pp.51-57.

Gliedman, John. "The Great Einstein Debate": *Omega Science Digest.* November/December 1983, pp.72-77 & 108-109.

Gliedman, John. "The Man Who Turns Einstein Upside Down": *Omega Science Digest.* January/Febraury 1985, pp.56-61.

Goldberg, Stanley. 1984, *Understanding Relativity: Origin and Impact of a Scientific Revolution.* USA: BirkhŠuser Boston, Inc.

Good, Timothy. 1997, *Beyond Top Secret: The Worldwide UFO Security Threat.* London: MacMillan Publishers Limited (Pan Books).

Grant, John. 1990, *The Great Unsolved Mysteries of Science.* New Jersey, USA: Chartwell Books (Book Sales, Inc).

Griffiths, David J. 1989, *Introduction to Electrodynamics.* 2nd Edition. New Jersey, USA: Prentice Hall Press.

Haish, Bernhard et al. "Beyond E=mc^2- A First Glimpse of a Universe Without Mass": *The Sciences* (New York Academy of Sciences). November/December 1994, Vol.34, No.6, pp.26-31.

Harris, Sidney. "What's So Funny about Hi-Tech?": *Omega Science Digest.* March/April 1985, pp.90-91.

Hawking, Stephen. 1988, *A Brief History of Time.* New York: Bantam Books.

Hawking, Stephen. 1996, *The Illustrated A Brief History of Time.* New York: Bantam Books.

Heisenberg, Werner. "Über den anschaulichen Inhalt der quantentheoretischen Kinematik und Mechanik" (translated as, "About the intuitive content of quantum theoretical kinematics and mechanics"): *Zeitschrift für Physik.* Volume 43 (1927), Issue 3-4, pp.172.198. Paper was received on March 23, 1927.

Hlavaty, Dr. Vaclav. 1957, *Geometry of Einstein's Unified Field Theory* (supported by the National Science Foundation in the U.S.). Groningen, Holland: P. Noordhoff Ltd.

Hlavaty, Dr. Vaclav. "The Elementary Basic Principles of the Unified Theory of Relativity": *Proceedings of the National Academy of Sciences.* USA, Vol.38, 1952, pp.343-347.

Hlavaty, Dr. Vaclav. "Report on the Recent Einstein Unified Field Theory": *Rendiconti del Seminario Matematico della Universiti di Padova. Anno XXIII*, 1954, pp.316-332.

Hoffman, Banesh & Dukas, Helen. 1972, *Albert Einstein: Creator and Rebel.* New York: Viking Press.

Huyghe, Patrick. "The Black Hole in Earth's Backyard": *Omega Science Digest.* January/February 1983, pp.14-17.

Infeld, Leopold. 1950, *Albert Einstein: His Work and its Influence on Our World.* New York: New York: John Wiley & Sons, Inc.

Izaks, Gerbrand J. & Westendorp, Rudi G.J. "Ill or just Old? Towards a conceptual framework of the relation between aging and disease": *BioMed Central Ltd.* BMC Geriatrics 2003, 3:7. December 19, 2003.

Janus, Owen. "Origins of 'Gospel of Jesus's Wife' Begins to Emerge": *Live Science.* August 24, 2015. http://www.livescience.com/51954-gospel-of-jesus-wife-origins.html

Jayawardhana, Ray. "The Age Paradox": *Astronomy* (American Publication). June 1993, pp.39-43.

Jeans, Sir James. 1951, *The Mathematical Theory of Electricity and Magnetism.* Fifth Edition. Cambridge University Press.

Kaufmann, William. "Was There a Big Bang?": *Omega Science Digest.* July/August 1982, p.116.

Kauffmann III, William. *What Makes Halton C. Arp the World's Most Controversial Astronomer: Omega Science Digest* (New York: The Hearst Corporation). January/February 1982, pp.74-77 & 124-127.

King, Karen L. "Jesus said to them, 'My wife...'"—A New Coptic Papyrus Fragment: *Harvard Theological Review.* April 1, 2014, Volume 107, Number 2, pp.131-159.

Laferty, Peter. 1990, *The Wayland Library of Science and Technology: The Universal Forces.* England: Wayland (Publishers) Limited.

Lanczos, Cornelius. 1965, *Albert Einstein and the Cosmic World Order.* New York: John Wiley & Sons, Inc.

Lawton, April. "The Moment of Creation": *Omega Science Digest.* September/October 1981, pp.36-41.

Levi-Civita, Prof. Tullio. 1949, *A Simplified Presentation of Einstein's Unified Field Theory.* London: Blackie & Son Limited.

Lester, Meera. 2005, Mary Magdalene—The Modern Guide to the Bible's Most Mysterious and Misunderstood Woman. Avon: England: F+W Publications, Inc.

McGraw-Hill Encyclopedia of Science and Technology. New York: McGraw-Hill Inc., 1987, 1992.

Mehra, Jagdish (editor). 1973, *The Physicist's Conception of Nature.* Dordrecht, Holland: D. Reidel Publishing Company.

Moltmann-Wendel, Elisabeth. 1982, *The Women Around Jesus.* New York: Crossroad Publishing Company.

Montagu, Ashley. "The Unknown Einstein": *Omega Science Digest* (New York: The Hearst Corporation). January/February 1986, pp.56-59 & 119.

Moore, Patrick & Nicholson, Iain. 1985, *The Universe.* London: William Collins Sons & Co Limited.

Moore, Ruth E. 1966, *Niels Bohr: The Man, His Science, and the World They Changed.* New York: Knopf.

Murphy, Pat & Doherty, Paul. 1996, *The Color of Nature.* San Francisco, USA: Chronicle Books.

Musgrave, Anne. "Realizing the Impossible Dream": *Omega Science Digest.* January/February 1981, pp.92-99.

Narlikar, Jayant V. & Padmanabhan, T. 1986, *Gravity, Gauge Theories and Quantum Cosmology.* Holland: D. Reidel Publishing Company.

Nipher, F. E. "Gravitational Repulsion": *Transactions of the Academy of Science of St. Louis.* Vol.23, No.5, November 8, 1917, pp.177-192.

Nipher, F. E. "New Evidence of a Relation Between Gravitational and Electrical Action and of Local Changes in the Electrical Potential of the Earth": *Transactions of the Academy of Science of St. Louis.* Vol.23, No.9, March 2, 1920, pp.383-387.

Okawa, Ryuho. "The Truth about Spiritual Possession". Published online on July 17, 2015, at http://info.happy-science.org/2015/127/.

Olenick, Richard P., Apostol, Tom M. & Goodstein, David L. 1985, *The Mechanical Universe: Introduction to Mechanics and Heat.* London:

Cambridge University Press.

Omega Science Digest. "Predictions-The Big Bang Paradox". May/June 1981, pp.102-103.

Omega Science Digest, "Predictions-Beacons of Deep Space". March/April 1982, p.108.

Omega Science Digest. "Predictions". July/August 1982, p.104.

Omega Science Digest. "Update: Low-Dose Radiation". November/December 1984, p.31.

Oppenheimer, Julius Robert. 1954, *Science and the Common Understanding.* London: Oxford University Press.

O'Neill, Amanda. 1993, *Gods & Demons—A Gallery of Unearthly Beings and Fantastic Beliefs.* London: Grange Books Limited.

Pais, Abraham. 1982, *Subtle is the Lord: The Science and the Life of Albert Einstein.* London: Oxford University Press.

Polkinghorne, J. C. 1990, *The Quantum World.* London: Penguin Books Limited.

Product Engineering. "Electrogravitics: Science or Daydream?" December 30, 1957, Volume 28, Number 26, p.12.

Rassam, Clive. "A Tale of Two Cultures": *New Scientist.* June 26, 1993, Number 1879, pp.30-33.

Robbins, Anthony. 1992, *Awaken the Giant Within.* New York: Simon & Schuster, p.185.

Roberts, Rev. Dr. Mark D. 2010, *What Language did Jesus Speak? Why Does It Matter?* http://www.patheos.com/blogs/markdroberts/series/what-language-did-jesus-speak-why-does-it-matter/.

Roland, Paul. 1995, *Revelations—The Wisdom of the Ages.* Sydney, Australia: Harper Collins Publishers.

Russell, Graham. 1972, *The Interpretation of Einstein's Unified Field Equations.* The University of Sydney (thesis).

Sabar, Ariel. "The Unbelievable Tale of Jesus's Wife": *The Atlantic.* July/August 2016. http://www.theatlantic.com/magazine/archive/2016/07/the-unbelievable-tale-of-jesus-wife/485573/

Schlumpf, Heidi. "Who framed Mary Magdalene?": U.S. Catholic. April 2000. Volume 65, Number 4, pp.12-16. Available online at *http://www.uscatholic.org/articles/200806/who-framed-mary-magdalene-27585.*

Seife, Charles. "Running on Empty": *New Scientist.* April 25, 1998, Volume 158, Number 2131, pp.36-37.

Skurzynski, Gloria. 1996, *Waves—The Electromagnetic Universe.* Washington, D.C.: The National Geographic Society.

Smith, Wilbert B. *Suggestions on Gravity Control Through Field Manipulation.* (unpublished), pp.1-3.

Spong, John Shelby. 1990, *Living in Sin? A Bishop Rethinks Human Sexuality.* New York: HarperCollins—Harper San Francisco.

Stemman, Roy. 1991, *Mysteries of the Universe.* London: Bloomsbury Books.

Tacey, David. 2003, *The Spirituality Revolution.* Sydney, Australia: Harper Collins Publishers.

Talbert, Ansel E. "Conquest of Gravity Aim of Top Scientists in U.S.": *New York Herald-Tribune.* November 20, 1955, pp.1 & 36.

Talmud Bavli, Tractate Sotah. Pocket Edition, 1723. Amsterdam: Moses Frankfurt.

The Encyclopedia Britannica, 15th Edition, 1987, Volume 24, pp.776-859.

The New Book of Popular Science. Deluxe Library Edition, 1991. New York: Grolier International, Inc.

The Encyclopedia of Space Travel and Astronomy 1979. London: Octopus Books Limited.

Tonnelat, M.A. 1966, *Einstein's Unified Field Theory.* New York, USA: Gordon and Breach Science Publishers, Inc.

Tralli, Nunzio. 1963, *Classical Electromagnetic Theory.* New York: McGraw-Hill Book Company, Inc.

Trefil, James. *The Accidental Universe: Omega Science Digest.* September/October 1984, pp.68-73.

Vardey, Lucinda (ed.). 1995, *God in All Worlds*. Alexandria, Australia: Millennium Books.

Verschoyle, W. D. "The Connection Between Gravity and Electricity": *English Mechanics*. May 8, 1936, pp.78-79.

Whyman, Kathryn. 1986, *Simply Science: Light and Lasers*. London: Franklin Watts / Aladdin Books (Gloucester Press) Ltd.

Whyman, Kathryn. 1989, *Hands on Science: Rainbows to Lasers*. London: Aladdin Books (Gloucester Press) Ltd.

Willis, Suzanne. "What is the premise of the book: The Final Theory?": *MadSci Network*. April 7, 2005.
http://www.madsci.org/posts/archives/2005-04/1112889896.Ph.r.html.

Witherington, Ben. *Mary, Mary, Extraordinary.*
http://www.beliefnet.com/Faiths/Christianity/2003/11/Mary-Mary-Extraordinary.aspx?p=2.

Wolf, Fred Alan. "Understanding the Quantum Leap": *Omega Science Digest* (New York: The Hearst Corporation). May/June 1982, pp.62-65 & 116.

Yam, Philip. "Exploiting Zero-Point Energy": *Scientific American*. December 1997, Volume 277, Number 6, pp.54-57.

Zemansky, Mark W., Sears, Francis W. & Young, Hugh D. 1982, *University Physics*. Sixth Edition. Addison-Wesley Publishing Company.

Zubrzycki, John. "Never Say Die: David Sinclair's Anti-Aging Quest": *The Sydney Morning Herald* (Good Weekend Magazine). October 5, 2015.

PICTURE CREDITS

Chapter 1
Albert Einstein: Photograph by Oren Jack Turner courtesy of the U.S. Library of Congress; Professor Stephen Hawking: Photograph by Kosala Bandara and available in Flickr.

Chapter 2
Albert Einstein aged 3 years: Unknown photographer taken in 1882 now in the public domain; James Clerk Maxwell: Photograph in the public domain https://commons.wikimedia.org/wiki/File:James_clerk_maxwell.jpg; Michael Faraday: Biblioteca de la Facultad de Derecho y Ciencias del Trabajo Universidad de Sevilla Flickr Image 106084; Atomic bomb test at Bikini Island: San Diego Air and Space Museum Archive Catalog # 10-0016036 (Flickr); Albert Einstein at the Patent Office: Photograph by Lucien Chavan (1868-1942); Sir Isaac Newton: Public domain image originally published in William Aubrey's *The National and Domestic History of England* (Volume 3), in 1878; Empedocles: Photograph by Skara Kommun (Flickr); Euclid statue outside the Oxford University Museum of Natural History: Photograph by Garrett Coakley (Flickr); Christiaan Huygens by Caspar Netscher, oil painting 1671, Museum Boerhaave, Leiden: Photograph by Rob Koopman (Flickr); Albert Einstein in his office at the University of Berlin in 1920: Unknown photographer published in *Scientific Monthly*; Einstein's house: Photograph by Edmond Chen (2007) and available from Flickr.

Chapter 3
Laser beam through convex lens: Photograph by Astroshots42 (Flickr); Crookes Radiometer: Public domain image taken in March 2005 from Wikipedia; Duke University: Photograph by Nan-Cheng Tsai; U.S. Navy officers conferring with Albert Einstein at his Princeton home in 1943: U.S. Naval Historical Center, Photo # 80-G-42919; Black hole: Modified image by SUNRISE from NASA (Flickr); NGC 4261 including image of black hole surrounded by torus of heated materials: NASA; Superconductivity: Photograph by sach1tb, (2006) from Flickr; Metal sphere art piece called *Heave*, photographed by Steven Lilley (2010) from Flickr; Jupiter: NASA; Photograph of Einstein with other famous scientific colleagues on November 30, 1931: Unknown photographer.

Chapter 4
Illustrations of Weightlessness, Gravity and Anti-gravity: SUNRISE; Dr. Vaclav Hlavaty: Unknown.

Chapter 5
Thomas Young: Portrait by Henry Perronet Briggs (1793–1844) in 1822; Thomas Young single and double slit experiment illustrations: SUNRISE; Arrival of electrons to form wave-like pattern in the double-slit experiment: Ethan Hein (Flickr); Car Traffic: Photograph by Mohamed Almari (Pexels); Richard Feynman: Photographed in the woods of the Robert Treat Paine Estate by Tamiko Thiel in 1984.

Chapter 6
Edwin Hubble: Wikipedia (unknown photographer thought to be owned by Western Washington University Planetarium but is no longer kept on record, so it is now presumed to be in the public domain); NGC1376: Hubble Heritage. Vera Rubin: Snapshot from *Most of our Universe is Missing;* COBE and WMAP images: NASA; Ripples in a pond: Christina Spiegeland free for commercial use; Bubbles: Image by Martin Fisch titled *Blister in the Sun*, 2010 (Flickr); Hubble Utra Deep Field photograph: NASA and ESA.

Chapter 7
Picture of the hands of 86-year-old grandmother Viboodhi Paati: Photograph by Vinoth Chandar (Flickr); DNA model: Photograph by ynse (Flickr); Scarlet jellyfish: Photographed by Bachware on May 23, 2016 (Wikipedia); Turtle in the ocean: Photo by Jeremy Bishop (Pexels); Two youthful girls Issa and Raleene Cabrera: Photograph by Vito Selma (Flickr).

Chapter 8
The face of Jesus Christ: Artist Warner Sallman (Flickr); The Ancient of Days (artist William Blake): AKG-Images Ltd; Solomon's Dream: Bridgeman Art Library.Ezekiel's Vision (artist Rafaello Sanzio): Public domain available from http://mail.wikipedia.org/pipermail/wikide-l/2005-April/012195.html; Jacobs Ladder (artist William Blake): British Museum; Statue of Zeus at Munich Glyptpotheke taken by Giovanni on April 12, 2006 (Flickr); Sculpture of Heraclitus at the Victoria and Albert Museum: Photograph by Afshin Darian (Flickr).

ABOUT THE AUTHOR

SUNRISE Information Services (SUNRISE) is a private research center aimed at producing original, stable, interesting and easy-to-read educational and research information for the global community, while uncovering new and original knowledge.

www.ingramcontent.com/pod-product-compliance
Lightning Source LLC
Chambersburg PA
CBHW040752220326
41597CB00029BA/4733